T0280226

Challenging Diversity

What challenges are presented by the claim that diversity should be celebrated? How should equality politics respond to controversial constituencies, such as smokers and recreational hunters, when they position themselves as disadvantaged? *Challenging Diversity* brings a new and original approach to key issues facing social, political and cultural theory. Critically engaging with feminist, radical democratic and liberal scholarship, the book addresses four major challenges confronting a radical equality politics, namely, what does equality mean for preferences and choices that appear harmful? are equality's subjects individuals, groups or something else? what power do dominant norms have to undermine equality-oriented reforms? and can radical practices endure when they collide with the mainstream? Taking examples from religion, gender, sexuality, state policy-making and intentional communities, *Challenging Diversity* maps new ways of understanding equality, explores the politics of its pursuit, and asks what kinds of diversity a radical version of equality engenders.

DAVINA COOPER is Professor of Law and Political Theory in the School of Law, University of Kent, and Director of the AHRB Research Centre in Law, Gender and Sexuality. Her previous publications include *Sexing the City: Lesbian and Gay Politics within the Activist State* (1994), *Power in Struggle: Feminism, Sexuality and the State* (1995), *Governing Out of Order: Space, Law and the Politics of Belonging* (1998).

Cambridge Cultural Social Studies

Series editors: JEFFREY C. ALEXANDER, *Department of Sociology, Yale University, and* STEVEN SEIDMAN, *Department of Sociology, University of Albany, State University of New York.*

Titles in the series

ILANA FRIEDRICH SILBER, *Virtuosity, Charisma, and Social Order* 0 521 41397 4 hardback

LINDA NICHOLSON AND STEVEN SEIDMAN (eds.), *Social Postmodernism* 0 521 47516 3 hardback 0 521 47571 6 paperback

WILLIAM BOGARD, *The Simulation of Surveillance* 0 521 55081 5 hardback 0 521 55561 2 paperback

SUZANNE R. KIRSCHNER, *The Religious and Romantic Origins of Psychoanalysis* 0 521 44401 2 hardback 0 521 55560 4 paperback

PAUL LICHTERMAN, *The Search for Political Community* 0 521 48286 0 hardback 0 521 48343 3 paperback

ROGER FRIEDLAND AND RICHARD HECHT, *To Rule Jerusalem* 0 521 44046 7 hardback

KENNETH H. TUCKER, JR., *French Revolutionary Syndicalism and the Public Sphere* 0 521 56359 3 hardback

ERIK RINGMAR, *Identity, Interest and Action* 0 521 56314 3 hardback

ALBERTO MELUCCI, *The Playing Self* 0 521 56401 8 hardback 0 521 56482 4 paperback

ALBERTO MELUCCI, *Challenging Codes* 0 521 57051 4 hardback 0 521 57843 4 paperback

SARAH M. CORSE, *Nationalism and Literature* 0 521 57002 6 hardback 0 521 57912 0 paperback

DARNELL M. HUNT, *Screening the Los Angeles 'Riots'* 0 521 57087 5 hardback 0 521 57814 0 paperback

(*list continues at end of book*)

Challenging Diversity

*Rethinking Equality and the
Value of Difference*

Davina Cooper

CAMBRIDGE
UNIVERSITY PRESS

CAMBRIDGE
UNIVERSITY PRESS

University Printing House, Cambridge CB2 8BS, United Kingdom

Cambridge University Press is part of the University of Cambridge.

It furthers the University's mission by disseminating knowledge in the pursuit of education, learning and research at the highest international levels of excellence.

www.cambridge.org
Information on this title: www.cambridge.org/9780521831833

© Davina Cooper 2004

First published 2004

A catalogue record for this publication is available from the British Library

ISBN 978-0-521-83183-3 Hardback
ISBN 978-0-521-53954-8 Paperback

For Didi

Contents

Acknowledgments   *page* ix

1   Introduction: mapping the terrain   1

2   Diversity politics: beyond a pluralism without limits   15

3   From blokes to smokes: theorising the difference   40

4   Towards equality of power   68

5   Normative encounters: the politics of same-sex spousal equality   91

6   Getting in the way: the social power of nuisance   118

7   Oppositional routines: the problem of embedding change   142

8   Safeguarding community pathways: 'possibly the happiest
    school in the world' and other porous places   165

9   Diversity through equality   191

Bibliography   208
Index   232

Acknowledgments

I began working on this book in 1997. It has evolved in many different forms, and some of the arguments explored here have been published in earlier incarnations elsewhere. Many people have helped me in working through my ideas. I want to thank Nicola Barker, Susan Boyd, Doris Buss, Jean Carabine, John Clarke, Joanne Conaghan, Kitty Cooper, Ray Cocks, Miriam David, Margaret Davies, Andy Dobson, Ruth Fletcher, Lieve Gies, Jon Goldberg-Hiller, Reg Greycar, Paddy Hillyard, Morris Kaplan, Susanne Karstedt, Liz Kingdom, Vivi Lachs, Maleiha Malik, Daniel Monk, Surya Monro, Janet Newman, Alan Norrie, Oliver Phillips, Helen Reece, Diane Richardson, Sally Sheldon, Margrit Shildrick, Joe Sim, Richard Sparks, Carl Stychin, Dania Thomas, Michael Thomson, Martin Wasik, Pnina Werbner and Claire Young. Many of the chapters have also been presented at conferences and in staff seminars; they have benefited enormously from the thoughtful and challenging comments received. I also want to thank the series editor, Steve Seidman, and the editor at Cambridge University Press, Sarah Caro, for their interest and assistance.

I wrote most of this book while a member of Keele Law Department. I am grateful to my colleagues there for their support, interest and encouragement, and for the practical assistance I received from the department's wonderful administrative staff. Elizabeth Corcoran, Nick Cartwright, Alba Bozo, Jenny Smith and Fadzai Kunaka at different times provided tremendous research assistance, and I want to mark my appreciation of the members of my Sociology of Law class 2003 who explored with me some of the arguments in the final chapters. The original field research on which this book draws comes from three different projects: 'Community, Democracy and the Governance of Difference' (1995–7), and 'Lesbian and Gay Policy-Making within Local Government 1990–2001' (2001–03), funded by the Economic and Social Research Council, and 'Governing Prefigurative Communities' (2001–03), funded by a grant from the Leverhulme Foundation.

Challenging Diversity could not have been written without the ongoing advice, support and intellectual questioning provided by Didi Herman, to whom I dedicate this book.

Earlier versions of some of the arguments developed in this book have appeared in the following published form: 'Against the current: Social pathways and the pursuit of enduring change', *Feminist Legal Studies* (2001); 'Like counting stars? Re-structuring equality and the socio-legal space of same-sex marriage', in R. Wintemute and M. Andenæs (eds.), *Legal Recognition of Same-Sex Partnerships* (2001); 'Promoting injury or freedom: Radical pluralism and Orthodox Jewish symbolism', *Ethnic and Racial Studies* (2000).

1

Introduction: mapping the terrain

Nonsmokers living in condominiums and apartment buildings who are bothered or made ill by drifting tobacco smoke can now fight back . . . These dangerous chemicals can . . . seep in above or below an apartment door, through poorly sealed walls, and in many other ways . . . More and more tenants are filing effective complaints – often with information provided by Action on Smoking and Health (ASH), the nation's oldest and largest antismoking organization.[1]

Are you involved in a dispute over custody, and your spouse smokes in the presence of the child . . . If so, you should read ASH's preliminary report on custody and smoking. In it you will learn that, in more than a dozen states, courts have ruled that whether or not a child is subjected to tobacco smoke is a factor which should be considered in deciding custody.[2]

In the 1990s, across several continents, battles over cigarette smoking took on a new momentum, as pro- and anti-smokers fought over institutional policy-making, local, state and federal laws, and the ethics of intervention and restraint. Tobacco advocates declared that smokers had become the new oppressed, smoking the new apartheid.[3] Discriminated against in job interviews and healthcare, banished from particular spaces and disproportionately taxed, smokers were on the receiving end of an intensifying attack and stigmatisation. As one tobacco advocate declared, 'We are at a crossroads. We as a people are beginning to see what is produced by state-sponsored intolerance of "target groups".'[4] Meanwhile, a second set of disputes raged over the management of social diversity. The disputes concerned the installation of religious symbols within publicly shared space. One particular battle, fought with considerable intensity in London, in the mid-1990s, involved the proposal to establish an eruv – a symbolic Jewish perimeter of connecting edifices, poles and wire. Stretched around the public space of largely urban neighbourhoods, eruvs work by converting the territory into a single private domain. Through doing this or-thodox Jews are enabled to carry on the Sabbath outside their home. Disputes over the establishment of eruvs occurred in the twentieth century across a num-ber of jurisdictions, revealing the complex and fraught relationship between identity, practice and symbolic meaning. In the United States, the American Civil Liberties Union (ACLU) became involved in challenging a New Jersey eruv on the grounds that it breached the wall between church and state.[5] In Britain, an attempt to build an eruv in north London generated formidable op-position among those who saw its creation as threatening prevailing norms of rationality, universality, the public/private divide and secular geographies (Cooper 1998a).

My third example moves from disputes over discrimination and the terms of belonging to tensions over equality, in this case the right to marry. The growing demand for, and implementation of, same-sex spousal recognition during this period demonstrates the global connections between modern social movements, as discourses of equal entitlement, legislative exemplars and shared activist tactics circulated through North America, Europe, Africa and Australasia. But spousal recognition has not been unequivocally celebrated, and opposition has not come only from right-wing forces. More radical commentators and activists have also rejected the drive to marry on the grounds that marriage maintains women's subordination, privatises welfare and unfairly privileges couples.

Smokers, the Jewish eruv, same-sex marriage: these three instances could be multiplied many times over, with other examples demonstrating similar, familiar tensions. Claims, for instance, by groups beyond the left's traditional gaze that they constitute a disadvantaged minority have emerged not just in relation to smoking, but also in relation to British recreational hunters, divorced fathers, and, in the United States, in relation to white, conservative Christians (Herman 1997; Smith 1998: 186). Similarly, paralleling the eruv conflict have been other disputes and schisms over the visibility and audibility of migrant ethnicities and minority faiths: male school students bearing Sikh daggers, Islamic girls wearing the hijab, the Muslim call to prayer across city spaces, and mass attendance at Hindu gatherings within residential areas (*Cherwell District Council*;[6] Galeotti 1993, 2002; Moruzzi 1994; Parekh 2000). Lastly, we can identify a host of other disputes, questioning the benefits of inclusion and accommodation. While these surfaced, with particular acuity, in the 1990s, in relation to lesbian and gay marriage, similar disputes can be seen in other areas. For instance, conflicts over the terms and nature of equality erupted in relation to the struggle for 'gay rights' in the military, as well as in relation to transgender politics, where the drive for respectability and passing among more conservative transsexuals rubbed up against the anti-conformist tendencies of a new transgender politics (Califia 1997).

Three challenges

These varied and contested narratives of discrimination, entitlement and accommodation motivate, navigate and illustrate my argument as it develops through the course of this book. The discussion that unfolds centres on the way in which values, collective identity and social structure intersect. In particular, I seek to explore how this intersection has been analysed and mobilised within those Western, left-wing frameworks that emerged in the wake of identity politics and poststructuralism. However, my discussion also goes beyond critique to propose an alternative analytical and normative account. To do so, I focus on three theoretical dilemmas and challenges. The first concerns the way in which we understand social relations. Gender, age, sexual orientation, race and disability have joined, and in some cases superseded, the left's traditional preoccupation with class. But what remains less clear are the criteria required for recognition as a social relation. What is needed before we consider particular forms of social treatment and practice to constitute something akin or analogous to gender or class? Do relations have to be broadly similar to those already recognised, assuming for the moment that these are in some way already alike? What would be required, for instance, before smoking could be recognised as social relations? Is evidence of discrimination, bias, or prejudice enough?

In *Justice and the Politics of Difference*, Iris Young (1990a) identifies five faces of oppression: exploitation, marginalisation, powerlessness, cultural imperialism and violence. Young (1990a: 64) argues that these faces 'function as criteria for determining whether individuals and groups are oppressed'. Although many groups will experience several of these categories combined in complex ways, '[t]he presence of any of these five . . . is sufficient.' Young's approach is useful in illuminating different aspects of oppression; however, for my purposes here, it is too broad. Because it constructs a typology out of the experiences of groups already defined as oppressed, it allows any group sharing a similarity with existing oppressed groups in some or other respect to slip in. At first glance, this may appear fair. But does it make sense to define constituencies, such as smokers or hunters, as being oppressed simply because they can bring themselves within one of Young's categories? In this book, I take a narrower approach. In relation to both celebrating diversity and thinking about how and why inequality matters, my starting point is the need to differentiate between the constituencies that present themselves as coming within the terms of disadvantage and oppression. How to do so, at both an analytical and normative level, is a primary challenge for this book.

The question of problematic identities, and their relationship to (in)equality, can also be approached from another angle: that of preferences or conduct. In contrast to the individualism of many liberal theorists, radical and multicultural scholars advocating equality (which not all of them do) have tended to associate equality claims with group membership. But how does this deal with those groups whose desires, choices or identities are perceived as socially harmful? While liberals might seek to detach injurious preferences and forms of conduct from an individual's equality entitlement, to what extent does this underestimate the impact of such preferences and actions on an individual's sense of personhood, particularly as it is read through the prism of group membership? How should an equality politics respond to harmful choices? And can we be confident, in the process, that our identification of harmful choices is a fair one? These questions raise the normative underside of equality, an underside that can become neglected when equality is reduced to formal parity in a way that ignores the consequences of particular choices; yet it is also an underside that risks becoming exaggerated and inflated when equality is tied too narrowly, or rendered too contingent, on the pursuit of 'benign' conduct.

In exploring the relationship of harm to equality, a further inversion is also required. What does equality mean in relation to dominant constituencies? If equality refers to something more complex than raising those with less, we might think about the application of equality to those subjectivities, preferences and conduct constituted through the terms of whiteness, heterosexuality, masculinity and economic power. Does the pursuit of equality mean powerful beings and doings are to be read as inherently injurious? And what is to be done to them as a result? Yet how to even frame this question, targeting as it does the status of the powerful, is a difficult one. Aside from the now well-made claim that people embrace contradictory, and complexly combined, social locations, is equality something that can be addressed through a model based on gender or racial or sexual groups, whether dominant or subordinate? What alternatives are there? How can equality be conceived in a way that recognises the relationship between being and doing, between social structure and individual agency, without losing the specificity of either? This is the book's second challenge.

The third challenge is a more concrete one. In recent years, certain progressive and radical writers have optimistically reassessed liberal discourse. Against left thinking which tended to privilege equality, seeing it as a good in its own right as well as necessary for the achievement of freedom (particularly positive freedom), writers such as Chantal Mouffe have turned to a more negative (anti-institutional) conception of liberty. While equality forms the central terrain for this book, my aim is to explore how it intersects other values, including that of freedom. In particular, I am interested in the 'normalisation' of values, and the implications this has for social relations. In other words, *Challenging Diversity* explores the ways in which values are embedded and materialised within the social, and the effects this has on the maintenance and contestation of inequality. My discussion pursues this along two routes. The first explores the ways in which prevailing norms reinforce existing inequalities and the impact this has on political reform; the second concerns the potential for transforming dominant norms, and embedding more radical or oppositional values. In relation to the latter, I ask: how can such values be sustained and reproduced? Can the normalisation of counter-hegemonic values and practices be safeguarded and rendered effective given the gravitational pull of dominant social processes? This is the third challenge.

The politics of diversity

My springboard into these challenges is the normative and analytical space of the politics of diversity. This space does not equate to an ideology or political stance. Rather, it is a broad, discursive space that emerged out of the very particular social, cultural and political conditions of the 1980s and 1990s – namely, the dismantling of the Soviet Union and of the communist regimes of eastern Europe, the upsurge of neo-liberal ideology, the backlash against radical feminism, the expansion of lesbian and gay politics, including the birth of Queer, and the struggles around multiculturalism and anti-racism. Intellectually, diversity politics sits at the confluence of several currents that include liberalism, communitarianism, poststructuralism, post-Marxism, feminism, post-colonialism and queer. Into the twenty-first century, the politics of diversity continue to exert a powerful influence on progressive and radical thinking in the West.

More generally, it is a space carved out of a particular reading of three broad, intersecting, political moments: the democratic, the right and the normative (see, e.g., Mouffe 2000). Within diversity politics, the legitimacy that emanates from majoritarian or mass politics represents an important foundation. However, it is one that has also been sullied by its twentieth-century history. Even leaving aside more extreme examples such as mid-century European fascism, the capacity of majoritarian politics to be rearticulated within right-wing populist terms has generated considerable anxiety. This anxiety has been fed by the contested mobilisations around race, gender and sexuality within the latter half of the twentieth century, and the attendant surge of attention within liberal states towards the needs and security of vulnerable minority practices and communities. In this context, the capacity of majoritarian politics to safeguard minority interests seems uncertain. Will Kymlicka (1995: 109), for instance, refers to the dangers cultural minorities face from democratic legislative structures where they risk being repeatedly outvoted on the resource issues and policies that are crucial to them (see also Kymlicka and Norman 2000: 9). While minority rights activists have continued to seek redress, protection, autonomy and equality from parliamentary legislative processes, they have also turned to non-majoritarian structures, most particularly the courts and media, often in the face of scepticism and hostility from progressive critics wedded to majoritarian routes.

The deployment of non-majoritarian tools, particularly of rights and entitlement, to defend and affirm subjugated identities has, however, in turn come under threat. For this is also a tale of conservative appropriation, in which right-wing forces colonise and inhabit the discourses progressive and left-wing constituencies have legitimated (see also Smith 1998: 177–9). In this case, it is the language of minority status and protection. While legal rights have traditionally been deployed to protect elite interests, especially property ones, what has been evident in more recent years is the collective self-interpellation of seemingly powerful groups through discourses of vulnerability. Thus, fathers' rights groups, recreational hunters and smokers, at the turn of the twenty-first century, appropriate and deploy the political weaponry of their opponents. If social, political and legal marginalisation provide a rationale for better treatment then they are marginalised, they are oppressed. And indeed, their use of these discourses is not entirely cynical; many conservative minorities do seem truly to feel their disenfranchisement, stuffed into the closets of liberal society that others so recently have vacated.

While for their opponents considerable scepticism regarding these claims remains, to the extent that a valorised and protected discursive space has been crafted around the experiences of minority status, it becomes hard to restrict group status to 'true' minorities without some clear basis for doing so. Or rather, many advocates of diversity politics have, by default, done just that, ignoring the contradictions and difficulties their arguments throw up. But 'false' minorities cannot be so easily erased from a terrain that has given them the key to their own recuperation. As a result, the politics of diversity has been forced to confront its own rather thin and hesitant processes of differentiation, a point that Chantal Mouffe (1996a: 246–7) recognises:

it is important to recognize the limits to pluralism which are required by a democratic politics that aims at challenging a wide range of relations of subordination. It is therefore necessary to distinguish the position I am defending here from the type of extreme pluralism that emphasises heterogeneity and incommensurability and according to which pluralism – understood as valorization of all differences – should have no limit.

Two main ways exist out of this dilemma. The first stays within the politics of minority entitlement, and works through the construction of criteria for identifying true minorities. In other words, imposters are uncovered by demonstrating that they, for example, do not lack power. Nancy Fraser (1997: 185) adopts a version of this first approach when she argues that questions of difference must not be divorced from 'material inequality, power differentials . . . and systemic relations of dominance and subordination' (see also Phelan 1994). However, for reasons explored in more detail in the chapters that follow, many proponents have preferred the second route, which distinguishes between minorities on the grounds of the value commentators ascribe to their conduct, norms and choices. This 'rediscovery of value' combines post-Marxism with post-postmodernism to argue for the need to recentre ethics and morality in evaluating 'legitimately different ways of being' (Weeks 1995: 11; see also Squires 1993). From this perspective, the quality of difference does not lie in the degree of oppression faced, but in the values anchoring its practices.

The articulation of these different moments, driven and refracted by the different intellectual and normative currents identified above, has rendered the space of diversity politics a messy and uncontainable one, leaking out in several directions. Nevertheless, diversity politics does coalesce around a series of shared concerns, questions and assumptions. First, it goes beyond the conditional liberal promise bestowed upon minorities of toleration, providing their differences are kept from affecting others. As Weeks (1993: 206–7) argues, toleration, with its pragmatic acceptance of unwelcomed difference, is not enough. The space of diversity politics is one in which social diversity is valued and celebrated, not just within the private sphere but within public life and fora as well. In large part this is because diversity is seen as enriching society and making choice possible (e.g., Kymlicka 1995: 84). In addition, for more radical proponents, diversity is celebrated for its capacity to challenge disciplinary conventions and the status quo. Yet, as I discuss in the chapter that follows, the charge to celebrate difference also brings difficult questions in its wake. While liberal multiculturalism has become absorbed, some might argue overabsorbed, with the question of 'illiberal' minorities, the politics of diversity has produced more equivocal responses to the question of conservative differences, particularly when expressed by non-dominant minorities. For the most part, this issue has been ignored by those whose interest is in affirming cultural and social pluralism. While some writers, such as Mouffe and Weeks, have turned to the need for limits in their drive to avoid a thoroughgoing libertarianism, confronting the challenge of reactionary or disagreeable forms of difference has tended to remain at the level of general statement rather than detail.

The second theme common to the space of diversity politics is a recognition of the role of power in structuring social diversity. This has several aspects. It acknowledges and takes into account the part played by gender, race and class, in particular in creating unequal social constituencies. It also encompasses an analysis and critique of the part played by dominant institutions and systems, most particularly the state and the economy in producing and maintaining relations of domination. Yet, diversity politics is not solely oriented towards a model of 'power over'. It also encompasses a recognition of the part played by social movements in struggling for change, drawing on the different forms of power they can access and deploy. The space of diversity politics, in different ways, sees the social and political as sites upon which to act in pursuit of change. In this sense, it differs from liberal multiculturalism, which tends to restrict its political aspirations to the acceptance and accommodation of existing forms of difference.

However, the changes diversity politics aspires to remain far less clear-cut. Diversity politics is anchored in the presumption that social and cultural differences should not be hierarchically organised. It is also a space in which freedom and privacy are defined as primary goals. However, the place of equality is more problematic. While some writers, such as Mouffe, identify equality as a central though not pre-eminent value, others have rejected equality on the grounds that it is too grounded in sameness to offer a useful framework for supporting social diversity (e.g. Flax 1992). Equality is not the only source of disagreement and tension, as I discuss in more detail in the chapter that follows. The space of diversity politics raises questions and embraces diverging opinions about the desired place of collective identities within society: are such identities oppressive or enabling? Are gender, class and race characteristics to bury or reform? And, more abstractly, is a new hegemony worth seeking or are all hegemonies, however radical they appear in theory, disastrous to the pursuit and maintenance of a freer, more enabling society?

In the final part of this chapter, I map out my analysis and argument in the chapters that follow. However, before turning to do so, I want briefly to clarify two sets of terminological usage. First, I have chosen to use the terms 'difference' and 'diversity' in their overlapping and everyday meaning, in contrast to those writers and commentators who use the terms in particular and distinct ways. While 'diversity' is often used to refer to cultural manifestations of 'natural' social variation, such as ethnic groups, 'difference' signals the operation of socially structured asymmetries, such as gender and race, and the identity positions articulated through them. More specifically, the term 'difference' has been deployed to highlight the binary form social inequalities can take, and to flag up the process by which the subordinate term, such as woman or the feminine, 'acts as a "dangerous supplement"... Each member of the pair refers to the other at the same time as it suppresses it. Thus the self-sufficiency of the more powerful side of the dichotomy is fragile' (Lacey 1998: 80). My reason for not adopting this distinction relates to the ways in which the two terms intersect. While scholarship on difference sometimes ends up dismissing diversity as a distortion, and discussion of diversity can obscure the power relations and epistemological concerns that difference highlights, my more everyday usage is intended to highlight the relationship, but also the overlap, between difference as a term of critique and diversity as a normative politics.

The second set of terms I want briefly to flag up concern the descriptors of political intelligibility. This has become a sticky issue in recent years on the left as each label has, in turn, undergone critique. While terms such as 'progressive' and 'radical' cannot be used innocently, avoiding them creates the problem of how to talk about left-wing politics. Although my focus here is on re-envisioning equality and the undoing of social relations, simply to refer to it as a politics of equality is to construct a framework which is too shallow and narrow, in that it marginalises other elements often associated with left-wing politics, such as environmentalism, community and an ethics of care. Without doubt, there is good reason to spell out, rather than assume, the character of a particular version of left politics. Nevertheless, I will use 'progressive' to designate a broadly social-democratic, welfare-oriented politics and 'radical' to indicate a politics predicated on the need to transform key institutions, structures and practices in pursuit of counter-norms sutured to a more equal or, to the extent this differs, a less hierarchical society. Such brief, thumbnail definitions raise more questions than they answer. However, my main aim here is simply to flag up that, despite the ontological assumptions about society embedded in both terms, I will use them in this book to aid my exploration of diversity and equality's place within a broad left politics.

Chapter outlines

The relationship between equality and diversity lies at the heart of the book. More specifically, I draw on the politics of diversity and my critical analysis of its claims as a springboard into a set of debates and questions. These concern the character of social disadvantage and its relationship to structured inequality; how equality might be conceptualised and pursued; the politics of social norms; and the struggle to protect and enhance counter-normative ways of being and living. I begin, in the chapter that follows, with the task of establishing, in greater detail, the remit and purchase of diversity politics as an analytical and normative discursive space. To do so, I draw on a conflict over religious symbolism that I signposted earlier – the controversy that erupted, in the early 1990s, over the proposed establishment of an eruv in London. Exploring this struggle provides a useful focus for several reasons. While affirming and celebrating cultural forms of difference represent, in many ways, the bedrock of diversity politics, the relationship of this celebratory stance to more conservative and traditional identities has remained underexplored, despite the increased attention paid to 'troublesome' minorities by multicultural and liberal writers. Orthodox Judaism provides an interesting site from which to consider these issues because it combines traditional religious beliefs and practices with a collective history of social disadvantage, exploitation and cultural marginalisation. My discussion offers a route into analysing the cultural and social politics surrounding the eruv; however, my primary objective is to use the eruv as a lens through which to consider diversity politics close up. In particular, I seek to highlight some of its key contradictions and dilemmas in relation to questions of identity, freedom, harm and the private domain.

Having mapped key elements of diversity politics, and having highlighted points of uncertainty, ambiguity and difficulty, chapter 3 goes on to address a major failing of much scholarship and analysis within this field: its inadequate conception of the social. Diversity politics asks us to affirm and celebrate difference, with its implication that difference equates to disadvantaged and vulnerable minority statuses. But without a conception of the social and how relations of inequality operate, how do we know which forms of disadvantage represent legitimate policy decisions, for instance, and which represent unacceptable oppressions? Where, for instance, does the negative treatment encountered in many polities by smokers fit within this schema? In order to provide a framework in which legitimate forms of policy differentiation may be distinguished from social inequality, the chapter explores two strands of analysis, both drawn from feminist writing in the field over the past two decades. The first focuses on systems and axes of oppression, the second on subordinate group membership. Having critically explored the benefits and disadvantages of both, I argue that neither are sufficient. The chapter then goes on to carve out an alternative approach to conceptualising inequality. Anchored in structural asymmetries of power, principles of inequality also encompass two other dimensions. The first concerns the ways in which asymmetries circulate through social institutions as a result of the permeation of norms and values to which they are articulated. The second concerns the distinctive and constitutive relationship of principles of inequality to social dynamics, such as capitalism and the intimate/impersonal.

To illustrate the approach adopted, the final section of the chapter considers the status of tobacco smoking. In particular, I consider the claim of some members of the smoking lobby that smoking has become the new 'apartheid', creating divisions and relationships analogous to gender, race or class. While I disagree with this claim, I want to suggest that the case of smoking is more complex than it might first appear. Its consideration therefore offers a pathway into thinking more deeply about the social character of inequality: namely, what is required before unpopular, stigmatised practices come to constitute asymmetrical social relationships, and what are the implications of being so defined?

The detour chapter 3 offers into social theory is intended to establish the theoretical foundations for my return, in chapter 4, to the political question of equality. Although equality is a central term of progressive political theory, its status within diversity politics has proved far more equivocal. One aim of this book is to recentre equality within radical theory by demonstrating how it can avoid the problems diversity proponents raise. In this chapter, I focus on three aspects of equality-talk: equality of 'who', 'what' and 'how'. In relation to these three elements, I argue for an approach which places the individual centre stage as equality's subject, treats power as equality's object, and sees undoing social relations of power as equality's strategy. At the same time, my aim is not to present equality as a settled utopia but as a strategy for present engagement. From this perspective, individual equality of power is an unattainable aspiration; however, my argument is that this does not matter. Unrealisable political ambitions are important in providing impetus and focus; they also provide a rationale for more concrete and situated engagements, namely, in this case, unpicking and dismantling relations of inequality.

In the final part of the chapter, I draw on a case-study to explore further the issue of how we 'undo' inequality and its implications for undoing categories of difference. My focus is the difficulties that arise when views as to the composition of particular inequalities conflict. The example I draw upon addresses the struggle between radical feminist and transgender activists over the character of the 'gender problem'. While radical feminists have argued that gender is organised around an asymmetrical binary of male and female, masculine and feminine, transgender activists have reconceptualised gender to emphasise the divide between 'authentic', biologically driven dimorphic genders and a subjugated and despised plurality of gendered locations. How gender inequality is interpreted will affect the strategies mobilised to unpick it. At the same time, I suggest, these two approaches do not have to be treated as antithetical. Common ground can be found which recognises and responds to gender inequality in both its forms.

At the heart of diversity politics, alongside a commitment to dismantling domination, is the celebration of social pluralism. Yet diversity politics, with its tendency to be weighted towards either power or value, has not done enough to enrich our understanding of the relationship between the two. To the extent that they have considered the relationship between values, writers such as Mouffe emphasise the tension and need to balance so that no one value monopolises the social. What, in particular, is missed in this process is the complex ways in which dominant values, constituted and embedded as norms, consolidate existing social asymmetries, and the capacity of equality to constitute, and to be articulated and read through, other normalised values. In chapter 5 I explore the power and effects of prevailing norms through a study of lesbian and gay spousal recognition. To develop my argument, I introduce the concept of normative organising principles. These principles identify ways of organising and reading the social through terms such as 'legitimacy', 'democracy', 'property' and the 'public/private'. In their dominant form, normative organising principles work to condense the space between description and social vision. How society is constitutes, with minor revisions, how it should be. Normative principles work both to define the good society and to protect it. Their complex circuits of referentiality bestow upon themselves, and the status quo they constitute and protect, enormous authority. However, although secure at a basic level since challenges tend to be re-coded as ridiculous, outrageous or impossible, dominant principles do adapt and change, while new normative articulations emerge and become consolidated, as chapter 5 explores.

Same-sex marriage (and registered partnerships) have been 'sold' as ways of enabling lesbians and gay men to achieve greater parity with heterosexuals, but the pursuit of marriage is not without its political costs. These are particularly apparent when we step back from examining a single indicium of (in)equality to the wider question of undoing social inequalities more generally. We then have to ask what implications spousal recognition has for social relations such as class or race. In approaching this question, the chapter focuses on the mediating role played by two distinct and important normative principles: proper place and the public/private divide. If these principles, in their dominant form, help to consolidate existing social relations through the spatial and functional allocation of 'appropriate' conduct, people and norms, to what extent are they disrupted by same-sex spousal recognition? Does equal access to marriage offer the possibility of reconstructing proper place and the public/private in more radical ways, or will same-sex marriage become absorbed within, and read through, existing normative principles? In addressing these questions, I do not want to suggest that the answer is pregiven. I therefore argue that the normative effects of spousal law reform depend, at least in part, on the concrete ways in which lesbian and gay spousal recognition is advocated, institutionalised and inhabited.

My exploration of the relationship between equality, inequality and dominant normative principles continues in chapter 6. There, I use the prism of harm to explore one way in which dominant norms are secured and naturalised and the effects this has on social relations of inequality. Discourse and regulatory structures position harm as common-sensical – something about which there can be little disagreement. Yet what counts as harm, and to whom, has been and continues to be intensely contested. Notions of harm are underpinned by social values, values that are both consolidated and obscured through harm's naturalising qualities.

To explore these processes further, chapter 6 focuses on nuisance. While nuisance is seen as a kind of penumbra of harm within daily discourse, given its connotations of irritation and annoyance, in law it represents a distinctive form of civil wrong or injury. My interest in using nuisance as a prism through which to explore the consolidation of dominant normative principles is threefold. First, nuisance provides a powerful, common-sense discourse which dismisses and objectifies that which it constructs as diverting attention or being 'in the way'. As a result, deconstructing nuisance draws attention to specific normative principles, in particular freedom, proper place and responsibility. It also, importantly, flags up the relationship between such norms and, on the one hand, relations of inequality – especially gender, race and class – and, on the other, social dynamics, such as the intimate/impersonal and community boundary work. But, third, nuisance discourse opens up a set of counter-normative tactics. In particular, it raises the possibility of being and causing a nuisance as a way of disrupting norms, such as proper place, or social dynamics, such as capitalism. It also offers a device for engaging in normative inversions through counter-intuitive nuisance utterances. In this way, claims of injury can provide the impetus to flipping over and rearticulating dominant normative principles, such as freedom and responsibility.

The challenge of securing new social norms and relations provides the focal point for chapters 7 and 8. My starting point for chapter 7 is the important political role played by the establishment of new practices. In other words, the pursuit of equality needs more than the right arguments and rhetoric. It also needs the establishment of new, sustainable, routinised practices. Securing new durable practices can be immensely difficult when they operate within a social landscape that renders them exhausting, foolish, hazardous or otherwise costly. Moreover, to the extent that such practices remain hard work, they are unlikely to become routinised or commonsense. Does this matter? Zygmunt Bauman (1993: 10–11) argues that moral practices must be grounded in ethical, non-rational choices which 'precede the consideration of purpose and the calculation of gains and losses'. My analysis takes quite a different approach. While it may be right for actions not to be considered moral if they are routinised or take the form of utilitarian evaluations, my argument is premised on both the importance and inevitability of routines in creating and underpinning counter-normative practices. To explore this further, my discussion in chapter 7 focuses on the processes of creating social pathways – strings of diachronically connected conduct – through de jure and de facto means. While the latter are established through being utilised and inhabited, de jure pathways are mapped and created externally. Yet, as I discuss, using as my example the development during the 1980s of British local government lesbian and gay equality policies, even officially constructed de jure pathways can encounter difficulties if they remain within an inhospitable environment.

One response to this problem is to try to change the environment to render it compatible with and supportive of new counter-normative practices. This strategy was deployed, with some success, by the British government of Margaret Thatcher in the late 1980s in order to compel public-sector professionals to adopt new market practices. Radical forces have, however, found it hard to exert a similar degree of environmental control. One context in which environments have been reorganised is that of the intentional or prefigurative community: the subject of chapter 8. Here, spatial, normative or other boundaries are installed to protect practices and community institutions from the gravitational pull of the wider environment. Yet, as I discuss, such boundaries tend to be highly porous. People, norms and laws enter; wider resource allocations exert their pull; and members exit. These processes place considerable demands on communities intent on maintaining counter-normative pathways. At the same time, boundary permeability is an important factor in many communities' survival, as well as in facilitating any broader impact they might have.

In the final chapter, I revisit, and knit together, *Challenging Diversity*'s main themes. As I have said, these revolve around the relationship between diversity and inequality, and the impact of other normative principles on equality's pursuit. Two questions anchor my closing discussion. First, what kinds of social diversity are embraced by a radical politics of equality? If such a politics entails undoing relations of inequality, what place, if any, is imagined for the cultural and social meanings currently assigned to gender and other asymmetries? Are these expected to disappear or can they outlast the inequalities to which they are presently attached? The second cluster of questions relate to the wider impact of prefigurative communities. If such communities can generate and sustain greater equality of power and reveal new ways in which equality can intersect other norms, penetrating wider society is vital. But what happens when community practices and critiques confront the status quo? Against communities' struggles for recognition and adaption, what counter-strategies do governmental and other institutional forces adopt? And what possibilities exist to thwart or overcome the marginalisation of alternative practices? It is with these questions that the book closes.

Notes

1. From ASH (Action on Smoking and Health) http://ash.org/smoking-in-condos-and-apartments.html.
2. From ASH, http://ash.org/custody-and-smoking.html.
3. E. Malcolm, 'We're victims of apartheid say smokers', *Daily Record*, 8 November 1999, http://no-smoking.org/nov99/11-09-99-4.html.
4. N. Kjono, 'More anti-tobacco violence', http://www.forces.org/writers/kjono/files/violenc3.htm.
5. *ACLU of New Jersey* v. *City of Long Branch* 670 F. Supp 1293 (DNJ 1987).
6. See *Cherwell District Council* and *Vadivale* 6 PAD (1991) 433, for dismissal of an appeal against Cherwell Council's refusal of permission to use an outhouse as a temple, on the grounds of disturbance to neighbours.

2

Diversity politics: beyond a pluralism without limits

The politics of diversity identifies a discursive space of considerable plasticity and range, produced from the particular admixture of post-Marxism, feminism, queer politics, liberalism and poststructuralism. In short, diversity politics names an analytical, critical and visionary space oriented around the constitutive role of power in producing social difference and identity, the expression and realisation of human agency, the value of diversity – both for individuals and society – the moral entitlement to freedom, the right to express difference in the public sphere and the disavowal of domination and oppression. For Jeffrey Weeks (1993: 208) it entails 'a value system . . . which accept[s] both diversity and a wider sense of human solidarity'. Zillah Eisenstein (1994: 175), focusing also on the normative dimension, writes:

I inscribe the liberal notion of privacy with an egalitarian text that does not assume sameness as a standard, but rather recognizes a radically pluralist individuality. Radical pluralism means differences are not ordered hierarchically; they are not set up as oppositions; they are not tied up with, or reflective of, power relations. They merely reflect diversity.

My aim in this chapter is to explore the common ground, and the lines of tension that circulate through the politics of diversity. To do so, I draw on the writing of a number of authors whose work is broadly situated within this terrain. The work of Chantal Mouffe, Jeffrey Weeks, Zillah Eisenstein and Judith Butler, albeit in quite different ways, has centrally addressed questions of social diversity, freedom and equality. At the same time, I want to avoid equating diversity politics with the claims of particular authors. Although certain writers' work has been key, the space of diversity politics is wider than this. More generally, it represents an influential terrain of thought in English-speaking post-industrial societies which has both shaped and been shaped by the numerous commentators and activists who operate within and across its borders.

To illuminate the common ground as well as the differences within diversity politics, my discussion focuses on a particular case-study. It concerns the conflict that took place through the 1990s over the establishment of London's first major eruv – a symbolic space enabling orthodox Jews to carry on the Sabbath beyond their homes.[1] Reading the eruv conflict through the lens of diversity politics highlights certain questions and issues. It also flags up certain ambiguities and evasions. The discussion that follows focuses on the following.[2] First, how are identities understood? what limits or boundaries are placed on the affirmation of diversity in general? and what implications does this have for a social constituency such as orthodox Jews? Second, what place does freedom occupy within diversity politics? does freedom depend on social diversity for its realisation or is freedom the means by which diversity comes into effect? Third, what role does privacy play within the normative politics of diversity? is privacy an entitlement? can it ever be a requirement? and what, if any, limitations should be imposed on its application? Finally, how does the politics of diversity understand harm? If harm is the primary brake on freedom, privacy and diversity, the values underscoring its identification need to be determined.

Eruv troubles

The eruv is an ancient Halakhic structure that allows orthodox Jews to carry (or push) objects outside the home on the Sabbath, an activity that would otherwise be forbidden (see Metzger 1989; Valins 2000). It functions by symbolically converting common areas into a single private domain through the construction of a continuous, enclosing boundary.[3] Although eruvs have existed for centuries, they became particularly prolific from the 1960s onwards, due, in part, to a growing religiosity and cultural confidence among observant Jews in the West (see Fishman 1995; Sharot 1991). The greatest number of eruvs exist in the United States, but others have also been installed in Israel, Australia and Canada. Although versatile in some respects, eruvs are also subject to exacting, complex rules (see Metzger 1989; Vincent and Warf 2002). These relate to the size of population and kinds of spaces that can be enclosed as well as the form acceptable boundaries or perimeters can take. The rigidity of eruv requirements became only too apparent in the proposal to build a London eruv in the early 1990s, when completion of the perimeter encircling the $6\frac{1}{2}$ square miles could not be completed without the installation of new poles and wire.[4]

While eruvs have been established in many cases with minimal public protest, others have generated opposition. For the most part, such opposition has revolved around the involvement of public bodies in the eruv's installation, whether in allowing their structures to become marked with the symbols of the eruv boundary or in permitting new structures to be situated. It is important to note that it is the *materialisation* of the eruv's symbolism which has triggered opposition, not the abstract idea of the eruv itself. In the United States the ACLU judicially challenged a New Jersey eruv in the 1980s, on the grounds it breached the separation of church and state.[5] The challenge was defeated in the court, which held that excessive entanglement did not exist, and the eruv was allowed to go ahead (see also Metzger 1989). Other cases have since followed in the United States and elsewhere.[6] In relation to the London eruv, the initial application for eighty poles to be linked by a thin high wire to complete the boundary became the focal point for an intense, unpredictably fraught, local struggle that caught Barnet council, residents, and a planning inquiry in its net. Ranged against the proponents of the eruv were a loosely assembled group of non-orthodox Jews, Christians and non-religious opponents. While opponents were diverse in terms of class, age and political affiliation, the leadership and momentum appeared, from interviews I carried out in the mid-1990s, to come largely from older, middle-class residents explicitly hostile to state-sponsored multiculturalism.

Opponents' objections were wide-ranging (see Cooper 1998a). They encompassed claims that the new poles would prove an unnecessary eyesore to, arguably, deeper and more fully felt social and cultural anxieties. In particular, critics argued, installing poles to demarcate an eruv constituted a territorial act that would generate antagonism and resentment; it would also produce 'harmful' demographic shifts as the area became reidentified (see also Lees 2003). What had been a modern, rational, largely secular neighbourhood would become, opponents feared, dominated by orthodox Jewish individuals, living – it was both suggested and implied – according to archaic, communalist norms, norms that would somehow be imposed on all local residents. This imposition distinguished the eruv from 'private' religious practices; unlike the latter, opponents argued, the eruv interior would embrace many who had not consented to living in orthodox Jewish space.

In February 1993 Barnet Council turned down the application for additional poles and wire on environmental grounds. United Synagogue, the formal applicants for planning permission, appealed. In November 1993, an inquiry was held, at the end of which the planning inspector recommended that permission be granted. This was accepted by the Minister for the Environment, and the council subsequently agreed to an amended version of the original application. However, this did not stop protestors, who, having unsuccessfully sought judicial review,[7] continued for many years to resist the eruv's installation. Indeed, it took until February 2003 for 'a traffic light symbol on a website . . . [to] be switched from amber to green'[8] indicating that the eruv was ready to be used.

On its surface, the eruv seems a straightforward issue for a politics of diversity committed to equality and freedom. Not only does it allow a historically subordinated community to mark territory in ways that acknowledge their presence within the land, but it also assists in sustaining minority traditions against countervailing assimilationist pressures. In addition, the eruv enables practical activities to take place that would otherwise remain forbidden, namely carrying on the Sabbath. This ability to carry and push is not a trivial matter. In addition to enabling house keys, handkerchiefs and food products to be borne, it also allows parents of young children to push a pram to synagogue or to family and friends' homes; similarly people in wheelchairs can be aided on the Sabbath where an eruv is in force. Consequently, supporters have described the eruv as an equal opportunities device: expanding freedom, above all, for mothers, and people with disabilities.

Yet the space of diversity politics also raises other concerns – potentially more troubling for those who see the eruv as a progressive structure. Does the affirmation of minorities include conservative groupings? And how is orthodox religion to be treated? While the eruv is not a technique for converting or proselytising outsiders, it has been described as a means of sustaining religious observance among Jews. Will its installation lead, then, to the desecularisation of neighbourhood spaces and the generation of less diverse communities? The eruv also raises questions about the position of 'outsiders' who may be affected but have not consented. If they see themselves as injured, does it follow that the eruv is harmful? To what extent do their views need to be considered in deciding whether an eruv can be installed?

Identity politics and the eruv

To address these questions, I want to begin by considering the place and status of cultural and social identity (see, e.g., Young 1990b: 122). Although there has been a shift away from the category of identity in certain intellectual quarters, the politics of diversity remains grounded in affirming and respecting vulnerable forms of difference (however the relationship between difference and identity is understood). Such affirmation distinguishes diversity politics from the more muted advocacy of tolerance apparent in traditional liberal frameworks. Jeffrey Weeks (1993: 206–7), for instance, expressly states that tolerance is not enough. For tolerance presumes a position of normative superiority – we tolerate things we do not like but believe have the right to exist. What diversity politics asks for is something stronger – affirming difference for its own sake and as a way of facilitating collective and individual freedom. This sense of sharedness and collective participation also distinguishes Weeks's approach from liberal individualism in a further respect. Radical tolerance involves more than a 'live and let live' philosophy. Weeks draws on Heller and Fehér (1988: 83) who argue for the need to recognise 'other people's alternative ways of life [as] *our* concern' – a theme I explore further in chapter 6. However, the level of responsibility Weeks reads into this, beyond the removal of obstacles, remains unclear.

A second way in which diversity politics departs from certain strands of liberalism concerns the latter's emphasis on the relative immutability or naturalness of certain forms of social difference. Diversity politics starts from the premise that all forms of variation are socially forged in the context of power, taking shape in and through antagonisms, reinventions, alliances and coalitions (see Mouffe 1996b: 24, 2000: 21; Weeks 1995: 31, 100). At the same time, in contrast to political movements that have sought to recreate and impose visions of the good individual, diversity politics does not see identity or social difference as a legitimate site on which outsiders can impose their dreams, telling people how they ought to be. This is perhaps unsurprising in relation to minority identities such as lesbian, gay or other 'queer' sexualities for example,[9] that have historically been subject to the destructive disciplinary intentions of others. However, one might expect a more critical and transformative gaze to have been applied to mainstream subjectivities and their potential for revision. In the main, this also has not happened. The space of diversity politics has produced a politics of personal and collective transformation, for instance in relation to transgender practices – but the idea that a radical politics might evaluate social subjectivities from the *outside* and declare them in need of 'treatment' has been substantially rejected. Hunt (1990: 24) maps out the dangers of doing otherwise in her critique of feminist 'purity' politics:

Speaking out, confronting our own and others' victimization . . . forms a crucial basis from which to critique the moral and normative basis of marriage, heterosexuality and male supremacy. But it must be embarked upon with care, because it can be, and is routinely, used to argue that most women are so psychologically brutalized that they cannot know their own interests and must have them defined for them by others.

How to think about and respond to difference and identity lies at the heart of the space carved out by the politics of diversity; however, this is an arena of considerable divergence. Different views exist regarding the kinds of identities or social locations to centre, as well as the strategic politics that surround them. I want to highlight here three overlapping approaches which I shall call transformative, transgressive and deconstructionist (the last two, in particular, highlight important differences between the version of diversity politics discussed here and liberal multiculturalism).

The starting point for the first approach is in the discarding of the privileged status afforded to the working class as the revolutionary subject, in order to recognise and validate the struggles of other social categories, including women, black people, ethnic minorities, lesbians and gay men, and people with disabilities. Proponents reject the notion of a single rupture or transformational struggle. As Heller and Fehér (1988: 33) declare, 'precisely because of the de-centred character of the social system, emancipatory actions need not focus on changing a single, all-encompassing and dominating centre or institution'. So women, for instance, are identified as struggling for equality rather than for the overthrow of patriarchy as a social totality. One risk of this approach is highlighted by Nancy Fraser (1997: 185) in her discussion of pluralist multiculturalism which, she argues, 'tends to substantialize identities, treating them as given positivities instead of as constructed relations'. While writers such as Chantal Mouffe (1996b: 24) acknowledge that identities are constituted out of power relations, there is nevertheless a tendency when constituencies are actually discussed, 'to balkanize culture, setting groups apart from one another' (Fraser 1997: 185). This is partly the result of not deploying a more structural analysis of the ways in which intersecting identities and locations (women, gay men, transgendered people, for instance) are produced. It is also the result of writing on coalitions and alliances, where stress is placed on political articulations between social constituencies – the 'chain of equivalence' (Mouffe 1992: 236). By focusing on the need to *create* connections, attention is deflected from any concrete analysis of the formative quality of existing connections and the constitutive role played by relations of power in their entangled development.

This first transformative approach is broadly characteristic of Mouffe's radical democracy. It differs quite substantially from the second I want briefly to refer to: that of sex radicalism. This latter adopts a more transgressive stance, to celebrate and embrace what it defines as the most marginalised, and denigrated, of identities: butch dykes, leathermen, sadomasochists, transsexuals. Going beyond the claim that differences should not be hierarchically organised (Eisenstein 1996: 184), sex radicals prioritise identities – and identities in this context largely mean sex/gender ones – according to their capacity to trouble regimes of the normal (see Cossman et al. 1997; Rubin 1989).[10]

The third approach, associated with the work of theorists such as Judith Butler (1990, 1993), treats identities not as liberatory collectivities, but as disciplinary fictions. While deconstructionist techniques have been used to highlight and sometimes affirm the silenced and excluded term (women, gay, etc.) that sustains its hegemonic counterpart, more radical deconstructionists remain equivocal about both the utility and plausibility of speaking from and mobilising less powerful identity categories. Judith Butler (1993: 227–8) writes:

As much as it is necessary to assert political demands through recourse to identity categories, and to lay claim to the power to name oneself . . . it is also impossible to sustain that kind of mastery over the trajectory of those categories within discourse . . . The expectation of self-determination that self-naming arouses is paradoxically contested by the historicity of the name itself.

Deconstructionist writing on gender and sexuality reveals the way in which identities such as woman or lesbian are based on false and distorting commonalities that work to regulate and constrain the agency of those who 'belong'. Equally seriously, attempts to uncover a shared essence and to secure a common ground operate as powerful and dangerous tools in the exclusion of others (in the case of feminist and sexual politics, transgendered women and bisexuals have been identified as the losers) (see also Butler 1993: 227). Yet, while deconstructionists demonstrate the errors of collective classifications, they have also been criticised for not identifying the political subjectivities with which identities might be replaced. Steven Seidman (1997: 134–5), for instance, suggests that

[p]oststructuralist gay theory edges beyond an anti-identity politics to a politics against identity *per se* . . . The latter seems driven by its centering on a politics of identity subversion and draws from romantic, antinomian, and anarchistic traditions for its cultural resonance . . . [T]heir focus on subverting identity seems to abstract from . . . institutional struggle and the social origin and efficacy of identity politics.

Diversity's subject

Despite the sometimes numbing, reiterated injunctions to celebrate or affirm difference and the other, apparent in different ways in all three approaches, the politics of diversity is not completely devoid of a more critical edge. Jeffrey Weeks (1995: 60), for instance, states that 'endorsement of diversity cannot stop at the boundaries of a culture'. In other words, diversity must also be facilitated by and within communities. Weeks draws on the arguments of Heller and Fehér (1988: 82), who distinguish between needs which entail oppression, domination, or sadistic or violent practices and those recognised as legitimate. Chantal Mouffe (1996c: 136) makes a similar argument.

There cannot be a pluralism which accepts *all* differences. We must be able to determine which differences should exist within a liberal democratic regime, because those differences are necessary for the realization of principles of liberty and equality . . . But necessarily, there are also differences which might exist but must be put into question, or should never be accepted, because these differences would create relations of subordination which are not acceptable within a pluralist democracy.

There are parallels here with some liberal multicultural writing which has focused, some would argue too hard and too obsessively, on the oppressive, illiberal or worthless qualities of certain minority constituencies and cultures (e.g., Okin 1999).

But what does the question of pluralism's limits mean for the context we are dealing with here: the desire for an eruv by orthodox British Jews? While we might say that modern orthodox Judaism lacks any *necessary* association with coercion or oppression in relation to members or outsiders, it does tend to be associated with more conservative social relations and cultural and moral positions than the usual celebrated subjects of diversity politics. As such, it reveals something of the slippage and ambiguity that operates when scholars come to draw limits. How the eruv is politically evaluated depends on what is taken as the relevant or defining difference, but this conditionality is rarely made explicit.

Three different, interconnected possibilities present themselves. The first focuses on the eruv as a discrete, meaning-bearing structure. The second looks more broadly at group norms and culture and asks whether such a group deserves support. Should state policies assist orthodox Jews in their attempts to make life easier for members, with the anticipated corollary that this will help to sustain their religious and cultural practices? The discursive terrain occupied by the politics of diversity tends to focus on one or other (and sometimes both in combination) of these first two possibilities. However, a third possibility also exists – one that I argue for in the rest of the book. This centres on the social relations of power within which a group and its structures are embedded: in this case, Britain and Europe's history of Christian hegemony and anti-Semitism (see for instance Cheyette 1995; Shapiro 1996). From the perspective of this third approach, the question to be addressed is not whether orthodox Jews constitute a beneficent or normatively neutral form of diversity; for if orthodox conduct, norms and values are seen as emerging from a history of persecution and violence, it would be unfair to engage in any evaluation without regard for how that history of persecution has informed both orthodox Jewish practices as well as the terms of evaluation within a Christian-based country. Instead, the central question to surface is whether installing an eruv would help to undo religious inequality; or, to put it another way, would denial of permission symbolically and practically reinforce Christianity's existing dominance?

This question is complicated by a further problem. For the most part, Jewish and Christian constituencies do not face each other across a single divide. Alliances or common interests between particular Jewish–Christian forces straddle the division between them, and on each side, acute disagreements and conflict exist. Many writers have commented on the ways in which group identities, including closely linked ones, emerge and are sustained through antagonism (see Connolly 1995: xvi; Fraser 1997: 185; Werbner 1996). This does not render diversity a zero-sum game, where to the extent to which one constituency or identity is strengthened, another is equally disempowered. At the same time, the notion that difference can be infinitely extended, with its vision of discrete, unconnected identity groups, is problematic. The reality of such intra-Jewish antagonisms was clearly evident in relation to the London eruv conflict (see Lees 2003). Intent on resisting the eruv's installation, many secular and non-orthodox Jews established coalitions with Christians and other opponents. This was not a 'false consciousness' on the part of Jewish critics – that is a failure to see that their 'true' interests lay with their orthodox kith. Rather, it illustrates the fact that developments beneficial to one Jewish constituency may be to the detriment of others. This puts in doubt the claim that installing an eruv or indeed any equivalent symbolic structure can support Jews as a whole struggling against Christian hegemony.

The complex character of freedom

In chapter 4, I return to the question of how to undo relations of inequality in contexts where views differ both as to the character of the inequality and to the strategies it implies. But here I want to turn to a different issue. The value diversity politics places on social difference is inextricably linked to its emphasis on freedom. However, the relationship between the two is a complex one. On the one hand, diversity can be seen as constitutive of freedom to the extent it makes it practically possible for people to choose and live out different kinds of lives. At the same time, social diversity functions as an expression, product and symbol of freedom – we read freedom's presence in the range of life choices that exist. But diversity also works to define freedom's subjects: the different social groups who pursue liberation through the lifting of restrictions, the abolition of exploitative or oppressive relations, and through accessing the resources necessary in order to flourish.

Despite freedom's centrality to the politics of diversity, it has received less detailed attention than one might expect from scholarship in this field in recent years. Authors repeatedly emphasise the importance of freedom, but the term remains underspecified in ways that throw into relief the careful and textured writing on freedom within critical liberal scholarship – a field that has productively worked through some of its discursive and philosophical antinomies on the subject (e.g., Kymlicka 1995; Waldron 1991; see also Bauman 1996). In the following discussion, I want to highlight three different understandings of freedom to have emerged within the space of diversity politics. These understandings do not map straightforwardly on to the three identity perspectives outlined above. While we might expect sex radicals to advocate negative freedom, radical democrats to advocate positive freedom and deconstructionists to talk about freedom as practice, the articulations are more complex and overlapping.

Despite the influence on diversity politics of Marxist and feminist writing, negative conceptions of freedom are particularly evident (e.g., Mouffe 1992: 228; Weeks 1993: 208). This contrasts with recent liberal scholarship which has taken a more critical stance towards the negative 'freedom from' (Galeotti 2002; Waldron 1991). At the same time, the attractiveness of negative freedom is not surprising given that diversity politics was largely forged in, and against, the shadow of Soviet communism and, more parochially, in response to the perceived excesses of radical feminism. As a result, both radical democrats such as Mouffe, and sex radicals, emphasise the need for liberty 'as the absence of impediments to the realization of our chosen ends' (Mouffe 1992: 228). In the light of individuals' capacity to exercise meaningful agency, freedom demands choice, and the withdrawal of the state, polity or social movement from the role of ever-watchful, authoritarian parent.

At the same time, the hard work performed by freedom within diversity politics requires an accompanying more positive conception, oriented towards the socioeconomic and cultural preconditions necessary for people to achieve their goals. Galeotti (2002) has written that this entails both assistance for individuals to counteract existing (or historic) forms of disadvantage, as well as supporting collective identities or social choices through targeted exemptions, entitlements and resources. While socialists have tended to focus on the importance of economic redistribution (and the reallocation of property rights) to positive forms of power, multicultural liberals have emphasised the importance of political and cultural support. Kymlicka (1995: 84), for instance, argues, 'For meaningful individual choice to be possible, individuals . . . also need access to a societal culture. Group-differentiated measures that secure and promote this access may, therefore, have a legitimate role to play in a liberal theory of justice' (though see Waldron 1995). From the perspective of positive freedom, social bonds and governmental practices are not primarily obstacles in the way of our freedom but rather are necessary to its achievement (e.g., Lacey 1998: 77). Thus, Jeffrey Weeks (1995: 65), in his discussion of the importance of choice, acknowledges, although does not develop, the social and institutional implications of the argument that 'if we recognize, as we must, that the individual can realize his or her freedom only with and through others, then the balance between our personal autonomy and our responsibility to others becomes the central issue'.

The third conception of freedom I want to highlight is somewhat different. It concerns freedom as a form of practice. Some suggest that transgression, with its ludic challenges to domination and 'the boundaries between discipline and indiscipline' (Dumm 1996: 117), might constitute such a practice.

Others, drawing on Foucault's later work, focus on freedom as practices of (and on) the self (e.g., Weeks 1995: 56–7, 69–70). Such practices may include transgressions; they may also include expressions of self-discipline and mastery. While these latter entail forms of conduct, they are principally concerned with an inner sense of self: specifically, the ways in which, through our actions, including through processes of denial and giving ourselves to a cause, community or faith, we perceive ourselves as free.

The contrasting implications of these different approaches to freedom can be seen if we consider them in the light of the eruv. At the most immediate level, the eruv is a manifestation of negative freedom as the prohibition on Sabbath carrying is lifted for an enlarged, but still bounded, area. As a result, religious injunctions to refrain from *schlepping* are pushed beyond the eruv boundary. In a sense, this form of 'freedom from' works to normalise orthodox Jews. With a relaxation of the constraints that apply only to them, they are more able to live like their non-orthodox neighbours. But does this necessarily extend their freedom? For instance, while orthodox mothers can make certain 'choices' where an eruv does not exist, such as staying home and having friends without young children over to visit rather than going to *shul*, where an eruv is in place these choices become compromised unless new valid reasons can be found.

The negative freedom engendered by the eruv – the removal of uneven constraints – may be a precondition of achieving greater formal equality. Yet, while advocates of diversity desire inequalities in impediments and prohibitions to disappear, they do not want to see a 'numbing' sameness or assimilation in its place. I do not want to overstate the claim that negative freedom generates homogeneity; nevertheless, freeing individuals from group-specific impediments may have isomorphic effects. Will Kymlicka (1995: 87–8) makes a similar point: '[A]s a culture is liberalized . . . the resulting cultural identity becomes thinner and less distinctive . . . people share less and less with their fellow members of the . . . group . . . and become more and more like the members of other nations'.

Sustaining difference may therefore also require a more positive conception of freedom, and the construction of an eruv, particularly when combined with state permission, can be seen as one small element in the exercise of positive freedom for orthodox Jews. Practically, the eruv makes it easier to be observant because it diminishes the associated costs. Symbolically, the practice and permission of installation signifies orthodox Jews' legitimate cultural presence within the community. From this perspective, the eruv is a technology of power in the sense of 'power to' rather than 'power over'; it enables Jews not simply to assimilate in the way provided for by the eruv's status as a negative form of freedom, but to also live out a different mode of existence.[11]

The third form of freedom identified, that of self-discipline or commitment, has a more complicated relationship to the eruv than the other two. While the eruv may directly facilitate freedom in its negative and positive guise, its relationship to practices of freedom is far more contingent. The absence of an eruv makes it possible to transgress Sabbath law in ways eliminated by the permissions an eruv brings forth. At the same time, the eruv creates new transgressions, for instance carrying or pushing beyond its boundary. Anxiety about the implications of this was expressed by one of the orthodox interviewees in Oliver Valins's (2000: 582) research on the proposal to establish an eruv in Manchester, in north-west England. This interviewee was not so much concerned with intentional transgression – unsurprising since compliance with religious law is a product of committed choice – but with the unwitting infractions that might follow relaxation of the law on Sabbath carrying.

You'd get a woman pushing a pram on *shabbas* [the Sabbath], there's nothing wrong with that as long as she doesn't go in the park and make a groove in the grass, or the child sitting in the park having a drink, and water spills on the grass which helps it to grow on *shabbas* . . . Then there's the side that people generally don't understand what you can carry, you see there'd be people that are not orthodox that'd maybe carry a wallet with money on *shabbas* (quoted in Valins 2000: 582–3).

Unwitting infractions may constitute new forms of conduct produced by the eruv, but they are not easily translated into acts of freedom. Acts of freedom, in this context, relate more to discipline, mastery and commitment, but here the eruv can be seen as a negative or at best irrelevant development. For many orthodox Jews, freedom comes from the capacity and strength to follow Halakhic law whatever the inconvenience caused. Indeed, where eruvs are established, some orthodox Jews continue to function *as if* an eruv were not present. For them, freedom comes from treating the Sabbath as special – as a time when carrying and pushing should be halted, even though such acts may be technically sanctioned through Jewish law.

Using the eruv as a lens through which to consider diversity politics' conception of freedom highlights the discrepant interpretations different ways of thinking about freedom generate. More generally, it raises questions about the conceptual and normative work we want freedom to do. The politics of diversity carves out a space in which freedom is presented as a good, but it is not always clear who it is good for and why. In particular, diversity politics seems to be agnostic as to whether the benefits of freedom lie in the breadth of an individual's choice, in their capacity to experience a sense of mastery, control or commitment, in attaining the ends they desire, living a fulfilling, worthwhile life (which may not be the same as the life they desire), or whether the benefits lie principally in the social richness that emanates from a flourishing, culturally diverse environment.

And what about the competing freedoms of others? If their liberties and autonomy have been reduced as a result of the eruv, is this something that needs to be taken into account? This claim was made in the battle over the London eruv; it was also made in the Canadian case of *Rosenberg* v. *City of Outremont*,[12] in which concerns were expressed that the eruv's erection 'involuntarily places non-members of the Orthodox Jewish faith within what amounts to a religious enclave with which they do not wish to be associated'. But does this constitute an infringement of their freedom? On what basis should such a determination be made? Diversity politics does not provide clear, uniform answers to these questions. What it does do, however, is to map two conceptual strands of thinking from which a response to these questions might be fashioned. The first of these concerns the issue of privacy.

The right to privacy

In the main, proponents of diversity concur in the importance of sustaining and strengthening private life. At the same time, as I discuss further in chapter 5, many are anxious to distinguish their notion of a private sphere from that pervading traditional liberal thought. Within the politics of diversity, the private is not usually identified as a literal place, such as the home. Instead, writers emphasise the importance of individuals being able to construct their own 'private' sphere – described by Jeffrey Weeks as the intimate aspects of our lives, 'the things that matter most directly to us as social beings' (Weeks 1995: 131), and by Iris Young (1990a: 119) as 'that aspect of his or her life and activity that any person has a right to exclude others from'. We can see this revalorisation of the private sphere, like the emphasis on freedom, as a reaction to the perceived hyper-politics of movements such as radical feminism, where every individual decision and action was deemed a legitimate object for account. At the centre of this hyper-politicisation, opponents claimed, was a normative monism: the assertion of a single right way of doing things. Thus, the insistence on a private domain is linked to the affirmation of diversity, and the insistence that there should be spheres of life free from the imperative to confess, justify and be judged (see also Phelan 1989: 131–2).

The concept of privacy is central to understanding the London eruv controversy. Not only does the eruv work by symbolically privatising community space, but proponents characterised it as a 'private' concern: of relevance only to them. As one leading figure I interviewed declared, 'Eventually we said we can't explain it or you'll never believe it... We presented it as a facility the community needs; to explain why we need it is our business. We just want you to respect the fact *we* understand it.' Hilton J makes a similar point in *Rosenberg* v. *Outremont* when he states that the eruv 'is only a religious zone for those who believe it to be one. That belief is limited to the practitioners of Orthodox Judaism.'[13] Opponents agreed that the London eruv was a privatising installation, but what they drew from this was quite different. For many, the private qualities of the eruv represented aggression rather than collective self-sufficiency. According to one objector interviewed, 'We saw the wire as an ethnocentric demonstration, the religious side is just a ruse . . . They put up poles as a demonstration of their territoriality – they don't need poles.' For opponents then, the claim that they should not 'see' the eruv – that it was invisible to them – failed on the issue of impact. In other words, because they were *affected* by the installation it could not be treated as a private device. At the same time, the very act of demarcating private space, symbolically performed by the eruv, constituted a major ground for their claim of harm.

What discursive and interrogatory space does the politics of diversity, then, provide in relation to the non-consensual *privatising* of public space? Presumably some limits need to be placed on which actions and structures can be walled off as out of bounds. However, identifying these limits is not usually made explicit within work affirming diversity. Part of the problem is the slippage in what privacy means, evident in the seemingly paradoxical reasoning of opponents that since the eruv privatises it cannot be treated as private. The physical appropriation of neighbourhood space presumably does not come within what Weeks and Young refer to as the intimate or legitimately exclusionary aspects of our lives. Symbolic privatisation may be treated more sympathetically since it allows for multiple overlapping privatisations by different forces or the privatisation of one group being mapped onto the more open or public spatial meanings of others. Yet, even a form of symbolic exclusion generated through the construction of an eruv boundary, and in the notional leasing of the land, raises issues different from those highlighted by the 'private' right to refrain from a duty to account.

The right not to justify is important for minorities, who are frequently required to explain their actions according to majority assumptions and norms in ways infrequently demanded of more powerful and naturalised conduct. A political entitlement to refrain from being held to account or opened up for scrutiny may therefore readjust local relations of power and protect constituencies that sometimes are forced to choose between offering disingenuous but acceptable rationales or honest ones that will mark them unfavourably by the majority. Yet, the right not to explain or be held to account is usually contingent on a prior finding that some activity or practice is *legitimately* private. This underscores the extent to which privacy claims are anchored in particular notions of value and, as I go on to discuss, in particular notions of harm.

The entitlement to privacy may have a rhetorical value, but it is doubtful whether it can act as an effective shield if the bases upon which it is claimed are not accepted. However, there is a second side to the way in which privacy is understood within diversity politics. With their emphasis on the right to *choose* privacy, diversity advocates also depart from liberals to the extent the latter *require* certain practices to remain private. For instance, in the case of the London eruv, some opponents criticised the planning application for thrusting a minority religion on outsiders through the use of shared outdoor spaces and, in some cases, individual homes to demarcate the boundary. Similarly, in the United States, liberal opponents objected to the state becoming entangled with religion through the act of giving permission for sections of the eruv boundary. These kinds of practices, it was suggested, took religion out of the private sphere to which it had been rightfully relegated. But within the politics of diversity, a different space is carved out in relation to community or state-mandated privacy. Iris Young (1990a: 120), for instance, declares: 'no persons, actions, or aspects of a person's life should be forced into privacy'. From this perspective, it would be hard to criticise the eruv for *failing* to function as a more discrete and invisible structure; for the private is a domain that is chosen, not imposed.

The power of harm

So far in this chapter I have explored three main bases according to which the construction of an eruv might garner a supportive response within the terms of diversity politics; these are the right to difference for orthodox Jews, individual and collective freedom, and privacy. At the same time, none of these grounds are absolute; each is circumscribed or subject to the brake of countervailing concerns. While this might take the form of competing stakes or norms, what I wish to explore further is when such conflicting interests, demands or preferences are expressed in terms of social harm or injury.

Within the discursive space of diversity politics, harm plays a central role. It provides the main brake on extending diversity and freedom ad infinitum in ways remarkably similar to the role of harm within liberalism. It also asserts a subject that is prior to the harm in a manner that conceptually undermines the social construction of subjectivity. In other words, harm assumes a subject that is already complete.[14] For the purposes of my discussion here, the question at the centre of what I want to address concerns how harm is conceived. The politics of diversity offers a broad spectrum in this regard. At one end of the continuum, harm refers to any action perceived by others as restricting their interests or personhood: 'that human interaction where free choice impinges on the freedom to choose of others' (Weeks 1995: 33). At the other, harm is restricted to a narrow, objective determination of hurt.

In considering how this interpretive spectrum operates in more detail, I want to explore the forms of harm identified by opponents in relation to the eruv, starting with physical damage. Opponents identified this as taking two forms: violence and demographic harm.[15] In the case of the former, it was argued, the eruv's establishment would not only provide a symbolic focus for vandalism, but, by identifying the Jewish community, render them an easy target for anti-Semitic action. Jewish opponents thus saw the eruv as personally endangering. According to one objector: 'A minority of the community having staked out and identified its precise territory leaves the whole Jewish community open to attack, abuse and vandalism.' The second form of physical harm identified by opponents concerned the demographic changes that the eruv's installation would produce, namely, an influx of orthodox Jews parallelling an equivalent outflow of others.[16] According to some, this would upset the demographic balance. Others suggested that it would make it impossible for non-orthodox Jews to feel that they still belonged or to fashion lives according to their own norms and values. These arguments echo claims made in the New Jersey dispute over the Tenafly eruv, where opponents expressed concerns about the impact on local politics and local shopkeeping of a substantial influx of orthodox Jews (Lees 2003).

Demographic shifts were also linked to a suggestion of economic appropriation. Opponents argued that the eruv would impact detrimentally on property prices,[17] although it was unclear quite what the effect would be: whether prices would rise within the eruv (and fall outside it), or generally decline as the eruv became seen as an orthodox Jewish ghetto (see also Lees 2003; Vincent and Warf 2002). Whichever way they went, price changes were deployed in argument to highlight the arbitrary, unpredictable and unfair consequences of the boundary-drawing process.

I have explored these fears and anxieties in more depth elsewhere (Cooper 1998a, 2001). I raise them here in order to help consider what kinds of effects are seen by diversity politics as harmful such as to militate against the eruv's construction or so as to require the consent of non-eruv-using residents. Clearly, the threat of physical violence can be seen as obstructing others' freedom, although it is unclear whether the eruv should take the 'blame' for this outcome. Similarly, the obligation to conform to orthodox standards of behaviour and dress – however unlikely this might prove to be – would also indicate coercion and constraints on freedom. But what about causing property values to fall? This is a form of interference with others' freedom, namely, the freedom that comes from alienation of land and property ownership. At the same time, it raises the question: do all obstructions to freedom constitute recognisable forms of harm within diversity politics? Or does a radical politics of diversity require not only the restriction of freedom, but also that the interest affected is recognised as having value? Yet who or what is to determine this?

Western legal regimes tend to privilege property ownership; for instance, compensation may be given to local homeowners for motorway building on the grounds that they have lost something of recognised value (a response not matched in the case of local ramblers, whose loss or injury is usually seen as far more amorphous, non-individualisable and intangible). A more progressive approach, in contrast, might be expected to prioritise other kinds of loss or harm, redefining which interests or activities are deemed to have worth. Such a reassignment, as I explore in chapter 6, is crucial if harm is to remain central without replaying traditional conceptions of value (see also Cooper 2001). The space of diversity politics embraces both subjective and objective perceptions of harm, that is according to people's own sense of injury or loss, on the one hand, and according to a defined list of recognised damage or injury, on the other. However, there has been little critical and theoretical exploration in any developed way of the relationship between harm and value.

The difficulty of conceptualising harm is at its most acute in relation to cultural or symbolic forms of injury. For opponents, the symbolism of the eruv – if their arguments are to be taken at face value – constituted its greatest danger. Objectors portrayed the eruv as 'a privatised Jewish space' that would impose a religious boundary upon secular people and, for Jewish immigrants, trigger memories of both shtetl and concentration camp.[18] The apparent suffering caused to all those within the eruv's vicinity from such memories echoes more general issues concerning a 'captive audience' (Greenawalt 1995: 86). Greenawalt uses this concept to explore offensive communications within the workplace, where people cannot simply walk away from symbols or speech acts they find injurious. Likewise, here, unless people relocated, they would be obliged to put up with a symbol read as disturbing or hurtful – the dilemma where expression occurs in 'non-private' spaces to which others have access or from which they cannot withdraw (see also MacKinnon 1993: 11).

Eggshell skulls

What is claimed here is public protection against offenses which undermine the stabilization of the public presence of different identities and their bearers.

(Galeotti 2002: 113)

The issue of symbolic and textually derived injuries has been extensively canvassed within feminist debates over pornography (e.g., Cossman et al. 1997; Lacombe 1994; Segal and McIntosh 1992), sexual fetishism (e.g., Butler 1997a; Valverde 1985; Vance 1989) and, to a lesser degree, hate speech (e.g., Butler 1997b; Matsuda 1993). In using the eruv as a lens through which to explore the approaches adopted towards symbolic harm, my focus is on the sceptical, but influential, stance – at least through the 1990s – of sex radicals and deconstructionists countering those voices confident about the damage caused by hate speech, explicit sexual imagery, and the negative portrayals of particular minorities.

For the most part, sex radicals and deconstructionists rejected an expansive conception of expressive or symbolic harm. In so doing, their primary concern was not with where the 'harm' took place or whether others were present – although some drew a distinction between public and private sites of injury; rather, they sought to challenge the claim that texts and symbols were injuryperforming. In discussing symbolic and textual harm in relation to the eruv, I want to focus on three claims: first, that the injury caused was not sufficiently serious to constitute harm; second, that the speech harm was mediated, and to some degree controlled, by the agency of recipients; and third, that meaning is constituted by context and history.

The first challenge to the claim 'expressions and symbols harm' relates to the degree of hurt caused by communicative acts. Somewhat ironically, given the emphasis placed on culture, deconstructionists and sex radicals have tended to treat the actual effects of cultural forms of expression as unknowable or as relatively insignificant, particularly in the face of accusations of injury. (Texts identified as destabilising or transgressive have tended to be granted more influence; see, e.g., Gotell 1997: 74, 76–7). During the feminist struggles of the 1980s in the United States, Britain and elsewhere over the effects of power-infused sexual imagery, sex radicals argued that those who disliked such images were offended rather than harmed (cf. MacKinnon 1993: 11). The reduction of harm to offence carried with it the presumption that the causer of offence was not under an obligation to mitigate or remove that which offended. If the (unintended) recipient had an 'eggshell skull' that was their own problem. Weeks (1995: 142), for instance, argues:

It is perfectly conceivable that these [sexual] explorations will offend countless people, but in an intensely plural society there appears to be no grounds for imposing one's own personal tastes on others if their actions can have no effect on you.

Weeks (1995: 151) makes a similar point in relation to debates over 'dress codes': the ultimately unsuccessful attempt in Britain during the 1980s to prohibit certain garments, such as black caps and dog collars, from lesbian and gay community venues, on the grounds of their fascist and slavery connotations.

In the end, the question of what you wear . . . whatever the wider issues, must be regarded as matters of cultural taste, however offensive that might be to some people. It becomes of public concern, however, when appearance becomes a weapon in, say, the advocacy of violence . . .

From this perspective, the harm caused to opponents reminded of concentration camps when they see the eruv perimeter is no more than offence. To be offended does not require the offender to mitigate the disturbance caused. Opponents are issued with the injunction: if you don't like the eruv, don't look at it. Opponents have no right to impose their eggshell skulls on others.

The possibility of disidentification – that opponents can choose to act differently – takes me to a second argument drawn from the work of Judith Butler. Butler's work starts from the well-established premise that texts (and symbols) do not possess a single interior meaning determined by authorial intention (Butler 1997b: 25). The text is, rather, a complex site constituted out of the different readings it makes possible. It therefore follows that, in the process of constructing (and not simply decoding) textual meaning, the recipient – along with the communities in which texts are produced, exchanged and consumed (Cossman et al. 1997: 25; Lacombe 1994: 38) – plays a central role.

Claims for the open-textured character of communication do not simply diminish the significance of audience's wounds; what is also being suggested, more fundamentally, is that the audience is not wounded. Judith Butler (1997a), for instance, interrogates words' injurious capability on the grounds that those who are 'hailed' have the capacity to challenge or refuse their interpellation. While Butler agrees that speech *can* harm, she claims, controversially, that hate speech has no necessary effect since it consists, effectively, of perlocutionary rather than illocutionary speech. Illocutionary speech performs its effects in the very act of doing; in other words speech functions as conduct – the conventional paradigmatic example being Austin's marital 'I pronounce you . . .' (see Butler 1993: 224–5). Consequently, if hate speech is illocutionary it harms as soon as it is produced regardless of the agency of the recipient. Mari Matsuda (1993: 24–5) broadly adopts this approach when she states, 'The negative effects of hate messages are real and immediate for the victims . . . As much as one may try to resist a piece of hate propaganda, the effect on one's self-esteem and sense of personal security is devastating.' Perlocutionary speech, on the other hand, is not conduct in the same way. While it may have consequences, these need to be demonstrated: saying and doing need to be distinguished so that it is possible to consider why and how speech injures (Butler 1997a: 102).

Applied to the eruv dispute, Butler's approach highlights the agency and responsibility of opponents – as well as their over-sensitivity – in generating the eruv's negative associations. Opponents' agency is such, Butler's work suggests, that they could rethink the eruv if they wished. Thus, rather than seeing them as the passive victims of a symbolically oppressive structure, her approach highlights their own investment in interpreting the eruv as a concentration camp or shtetl. At the same time, we might argue, this situation is different from one where people are recipients of hate speech, with the hypothetical capacity to reverse the meaning of names by which they are hailed: as dyke, queer, fag, Yid (see also Butler 1993: ch. 8). The intention of eruv proponents is not to impose a pernicious structure on others, but to create a beneficial installation; thus opponents' reading of it as harmful is already an inverted reading: a *mis*reading.

Given this, a more pertinent question may be whether secular Jews in fact have the capacity to read the eruv as a *valid* religious structure, or at least as an installation that will not symbolically injure them. Is this to ask opponents to be other than they are? While multiculturalists have tended to be more sympathetic to the eruv, recognising its value to orthodox Jews and perceiving it as personally neutral (even if not beneficial) to themselves, the interpretive context in which opponents find themselves is one of a normative commitment to mainstream, hegemonic universalism and liberal rationality. From this perspective, the eruv is pre-modern, ghettoising, religious mumbo-jumbo, and even if we do not see context as governing reading, it is hard to see how, without a substantial shift in outlook, opponents might read the eruv as anything else.

The normative and epistemological settledness within which recipients of communications operate is underestimated by Judith Butler. Although Butler (1997a: 50) is concerned that the development of American hate speech doctrine exaggerates the power of the addressor or utterer (see also Butler 1993: 227), she does not show the same concern about overestimating or decontextualising the interpretive power of the utteree. But if meanings predate the speaker, to what extent does the recipient of a communication possess the agency to recreate its meaning, and hence its effects, at the point of receipt?

The dilemma of recipients' agency, structured as it is by their social context, takes me to a third issue raised by the question of symbolic harm: the role played by spatial meaning in shaping context. In the case of the London eruv, one ground opponents drew upon to stress its inappropriateness was its installation on land through which meanings already circulated – in other words, the eruv's terrain was not culturally vacant soil. Although this point has applicability to the entirety of the eruv boundary, one particular instance, often quoted in the conflict, related to a part of the proposed eruv boundary initially mapped to cross a Church of England school entrance. Opponents claimed that this was particularly inappropriate; since the space was already Christian, it should not be traversed by a Jewish 'boundary line' (see Cooper 1998a).

Yet even if the space of diversity politics treats the cultural and social context as determining what the eruv means, does it follow from this that any injury to prevailing cultural meanings should be recognised and ameliorated? In the main, the politics of diversity has paid little attention to claims of cultural injury uttered by dominant social groups. At first glance, the reasons for this disregard are easily comprehensible in terms of equality. Majority groups are not entitled to retain more cultural status and recognition than others (see chapter 1); they must therefore let go of some of their privilege, however painful such a letting go may feel. However, for the most part, the rationale for disregarding majority claims tends to go unstated, and as a result has remained theoretically undeveloped. This causes problems when the division between dominant and minority groups is no longer clear-cut or is contested. As I discuss in the chapter that follows, when smokers or hunters say that they are subject to opprobrium and marginalisation, does this experience represent an attack on the symbolic and cultural power of dominant groups or more vulnerable ones?

Conclusion

My aim in this chapter has been to use the eruv conflict as a prism through which to explore the space of diversity politics. As I established at the outset, diversity politics is not a single, unified perspective but a discursive terrain organised around particular questions, premises and concerns. While the boundary between diversity politics and a progressive, multicultural liberalism may have thinned, perspectives such as radical pluralism, radical democracy and sex radicalism continue to draw upon (and feed) post-Marxist, feminist, anti-racist and queer writing, as well as building upon more mainstream liberal trajectories.

In addressing the question of how diversity politics might respond to the 'eruv wars', I have sought to highlight the different approaches taken as well as the common ground. My analysis suggests that despite some ambivalence towards the values and practices of orthodox Jews, diversity politics generally provides a set of premises and perspectives favourable to the eruv's installation, particularly in the absence of concrete harm to others. However, the main purpose of this discussion has not been to determine on which side diversity politics would 'come down', but to illuminate the values, assumptions and concerns which lie at its heart, since these provide the springboard for the rest of my discussion in *Challenging Diversity*.

My analysis has focused on four issues. The first concerns how we think about social identity. No doubt because this lies at the core of diversity politics, accounts of identity combine a normative valorisation with a high degree of analytical sophistication, particularly in poststructuralist-influenced work on identity formation, hybridity and the impact of power. At the same time, the commitment to social diversity on the one hand, and individual agency on the other, has generated considerable reluctance to interrogate identity 'choices'. This refusal to subject other people's identities to critique diverges not only from political stances such as radical and revolutionary feminism, but also from the work of progressive liberal writers such as Kymlicka (1995), who argue that illiberal minorities need assistance in order to change and become, implicitly, 'more like us'. At the same time, this has not meant the complete absence of all difference critique. For democratic pluralists such as Chantal Mouffe and Jeffrey Weeks, excluding authoritarian or exploitative forms of difference is a central element in their normative projects. According to Mouffe (1992: 13), 'criteria must exist to decide between what is admissible and what is not'.

In the main, however, the process of determining which differences 'count' has been undertheorised within the space diversity politics has opened up. While it is an uncomfortable form of accounting for some, with its explicit assertion that not all differences can be celebrated and supported and that choices must be made, it is also unavoidable. Liberal advocates of state neutrality may argue that public bodies can avoid making explicit choices between different forms of the good life, but even leaving aside the procedural choices that remain, societies and polities implicitly (as well as explicitly) make certain ways of life thinkable and unthinkable, possible and impossible, easier or more awkward to sustain. Political questions therefore emerge as to the bases, principles or criteria for such different responses. Should they be based on the respective value of different choices or their respective power? And how do these two intersect?

The second main strand of discussion in this chapter concerned freedom. Again, a commitment to freedom is at the heart of diversity politics, yet, as my exploration revealed, freedom is conceived of in quite different, if not inherently incompatible, ways. These range from freedom from interference to freedom as positive assistance and freedom as a practice of discipline or commitment. Differences in conceptualising freedom are not the only differences that operate. While freedom is generally assumed to be a 'good', the value of freedom is variously conceived. So, freedom is valued because it generates social diversity, choice, commitment, the good life, or the subjective sensation of feeling free. Yet, in the main, these differing rationales are implicit. In contrast to recent liberal scholarship which has worked through the opposition between community and individual freedom in detailed, nuanced ways, the more radical discursive space of diversity politics has tended to underspecify the role and meaning of freedom despite its normative and rhetorical centrality.

Individual freedom and the support of social forms of difference may lie at the heart of diversity politics, but two other terms also play an important supportive role. The first of these, discussed in this chapter, is privacy. Against the apparent emphasis on full disclosure within more authoritarian forms of socialist and feminist politics, valuing diversity is characterised by the stress placed on individuals' right, though not obligation, to privacy and to provide a selective form of accounting (and access). However, in adopting this stance, the discursive space of diversity politics has demonstrated less clarity about where the limits on privacy lie, for instance when privacy rights are exercised in discriminatory ways or to shore up economic inequalities. While the eruv example does not raise these particular privacy dilemmas, it does raise the question of how to tackle a situation where one side's right not to be held to account for something deemed only of relevance to themselves is another side's illegitimate appropriation – seizing and withdrawing something that belongs to 'all'.

The last concept explored in this chapter is that of harm. Liberal scholarship on difference and multiculturalism has developed a heightened level of scrutiny and anxiety regarding the 'illiberal' practices of minorities. This anxiety, albeit to a lesser extent, pervades more radical commentaries, detracting from their traditional concern with the oppressive and problematic character of more powerful social forces. But because the diversity politics discussed here has tended to focus on valorised minority constituencies or has taken a more theoretical or normative form, very little detailed discussion has been generated about how to deal with harmful or oppressive forms of difference.

More generally, writing within the field of diversity politics highlights harm's centrality, yet, with some exceptions, fails to unpack the values on which it is based. This does not mean that scholarship presents harm as valueless; as I discussed in this chapter, writers equate harm with the value-laden terms of domination, oppression and impediments to freedom. However, by failing to explore in more depth the normative choices harm conceptually requires, questions remain. These arise when utterances of harm or injury occur over existing property-based entitlements and in relation to the harm potentially caused by offensive speech or symbols. In the latter instance, the question raised is not only whether harm discourse should protect people's social sensibilities, feelings of belonging and entitlement to be free from mental distress but also evidentiary questions and standards of proof regarding how we know that such injuries have been caused, and who or what they have been caused by.

In the chapters that follow I explore the questions and issues outlined here in more depth. I start with the first set of concerns discussed, namely how we think about social difference in the context of inequality. My argument centres on the problematic I highlighted: on what basis do we differentiate between social constituencies? Arguing against an approach that centres exclusively on the relative value of different ways of being, I suggest that we concern ourselves instead with the question of social inequality. But such an approach does not deliver straightforward answers, as the claims of conservative 'minorities' make clear. I therefore argue that we need a theoretical framework which can differentiate between inequalities according to their constitutive place within the social. This requires the reinstallation of a far more structural understanding of inequality and power than the space of diversity politics has tended to support. But it also requires us to engage fully with the challenges and insights diversity politics poses. How we recentre social structure in a way that draws on, and responds to, these challenges and insights tests the quality of the framework offered, as I explore in the chapters that follow.

Notes

1. My discussion of the London eruv conflict is based on field research conducted between 1995–7, see Cooper (1998a).
2. One important issue beyond the parameters of this discussion concerns the question of the state or public domain's responsibility for dealing with conflicts such as this one. How to respond to conflicts over cultural and social differences has been explored by a number of writers working from different perspectives. I have also focused on these issues in more detail in earlier work. This book, however, is principally concerned with the social and normative theory underpinning such political and policy-oriented questions.
3. As Valins (2000: 581–2) notes, residents within the area then need to be 'ritually unified (as though they were a single household) by either collecting food from each of them and the appropriate blessing recited, or else symbolically renting the area from the authorities'.
4. United Synagogue Eruv Committee, Barnet Council Briefing, 3 June 1992.
5. *ACLU of New Jersey* v. *City of Long Branch* 670 F. Supp 1293 (DNJ 1987).
6. See for instance *Tenafly Eruv Association* v. *Borough of Tenafly* 309 F. 3d 144 (3d Cir. 2002); *Rosenberg* v. *City of Outremont* (2002), 84 CRR (2d) 331.
7. See 'Eruv objection thrown out', *Hampstead and Highgate Express*, 20 June 1997; 'Eruv survives legal challenge', *Jewish Chronicle*, 20 June 1997.
8. T. Kirby, 'New rules on observing Sabbath divide a London community', *Independent*, 28 Feb. 2003.
9. Although, to the extent that these identities also emerge as a result of the play of power, it could be argued that they too should be subject to a politics of critical questioning and transformation.
10. I have questioned elsewhere (Cooper 1995a: ch. 3) the extent to which these identities actually challenge the status quo.
11. There are echoes here of Kymlicka's (1995: 126) argument that 'individual choice is the range of options passed down to us by our culture. Deciding how to live our lives is, in the first instance, a matter of exploring the possibilities made available by our culture.' But does a more radical version of diversity require the freedom to

choose which community in which to participate (e.g. Weeks 1993: 208), to be able to move between different ways of living? While the eruv may not overtly hinder this latter process, it arguably does nothing to facilitate it; the positive freedom to live an observant Jewish way of life is largely aimed at existing Jews.

12. (2002), 84 CRR (2d) 331 at 336.
13. *Ibid.* at 343.
14. Even if subjectivity is identified as, at least in part, constituted by the 'harm', the notion of harm still assumes an anterior subjectivity in relation to who or what has been harmed.
15. See interviews on file with author; see also item 3, Town Planning and Research Committee, Barnet Council, 24 Feb. 1993; Barnet Eruv Objectors Group Statement, May 1995.
16. See item 3, para 8.2.10, Town Planning and Research Committee, Barnet Council, 24 Feb. 1993; see also Jacobs, Witness Statement, Inquiry Report, 30 Nov.–13 Dec. 1993.
17. See Letter of Objection, item 4, appendix B, Town Planning and Research Committee, Barnet Council, 27 Oct. 1993.
18. See *Barnet London Borough Council* and *United Synagogue Eruv Committee*, 10 PAD 209, para 4.57; see also objections, Inquiry Report, 30 Nov.–13 Dec. 1993.

3

From blokes to smokes: theorising the difference

Which identity claims are rooted in the defense of social relations of inequality and domination? And which are rooted in a challenge to such relations?... Which differences... should a democratic society seek to foster, and which, on the contrary, should it aim to abolish?   (Fraser 1997: 184)

The space of diversity politics, grounded in a recognition of multiple forms of social difference, vitalised by its concern for multiple forms of inequality, reverses the monolithic tendencies found within Marxism and radical feminism. Yet, in its enthusiasm to challenge disadvantage and to celebrate variety, diversity politics comes unstuck when it is confronted with less attractive ways of living and being. As I discussed in chapter 2, radical democrats and radical pluralists, such as Chantal Mouffe and Jeffrey Weeks, do not deny the need to draw limits in the differences worth supporting; however, their work gives little detailed or practical guidance on how these limits can be drawn. Practices such as oppression, subordination and limiting diversity sound like solid bases for identifying constituencies not worthy of support, yet even these terms can be 'subverted' in a pluralist model of power and discursively deployed against opponents by groups such as hunters, religious minorities and smokers.

My argument in this chapter is that we need a more detailed and theorised approach to thinking about different social constituencies. While writers within the field of diversity politics tend to constitute groups out of their beings and doings, the approach I adopt centres on their social location. In arguing for the importance of a more structural conception of power, inequality and the social, I want to draw on older influences, particularly that of feminist work of the latter half of the twentieth century. At the same time, I am keen that the framework to emerge integrates the pluralism and contingency shown by post-Marxist diversity politics.

I have chosen to focus on feminist writing for two primary reasons. First, it, more than most, has struggled with the question of how to recognise, and to conceptualise relations between, multiple forms of inequality. Second, it has developed a body of work which, in exploring how sites and institutions are gendered, racialised and sexualised, refuses to limit thinking about inequality to relations between subjects. My engagement with feminist work takes the following form. The first part of the chapter explores the ways in which social inequalities have been conceptualised within feminist writing since the early 1970s. In particular, I explore two paradigms: one treats oppressions or social relations as interlocking systems or axes, the other as overlapping group membership. Both the benefits and flaws of these two models having been identified, the second part of the chapter goes on to suggest an alternative approach, based on the concept of organising principles. This term has been used in different ways by different writers; I use it here to identify those social inequalities that shape, but can also be read off from, the configurations different societies take. The final part of the chapter draws on the framework so far developed to consider one particular candidate for status as an organising principle of inequality: smoking.

Cigarette smoking is, in my view, interesting because of the way in which rights-based groups have mobilised to defend smoking and to confront what they see as cigarette users' increasingly entrenched position as an oppressed or disadvantaged minority. However, to ask whether smoking constitutes an organising principle of inequality, as opposed to simply being a prohibited or unpopular practice, raises a more general issue. Why do we need a deeper theorisation of principles of inequality: to know or determine which characteristics count? Is it not enough to say that smokers are, in certain contexts, disadvantaged? What purpose is served in deciding whether smoking and other disadvantaged activities constitute relations of inequality analogous to gender, race or class? Indeed, does not even asking the question fall into the trap of equivalence that Christian right activists, recreational hunters and tobacco pressure groups seek to establish? In other words, the very process of considering cigarette smoking alongside class, gender, race and sexuality suggests a framework able to incorporate any grouping that can come within its terms. This seems to trivialise recognised inequalities; amplifying their similarities may also deflect attention from their historical and current differences. At the same time, a refusal to think about inequality categorically will not stop prevailing tendencies to homologise – enlisting differences in pursuit of a picture of disadvantage which seeks to be comprehensive at the very moment when its 'etcetera' reminds us that it cannot be. It will also not stop the rhetorical claims to injustices suffered by conservative social forces.

In chapter 4, I argue for an equality politics that seeks to undo relations of inequality: in other words this is an equality politics intent on converting asymmetries of power into socially indifferent or evacuated distinctions. But this does not mean that all social distinctions, even unequal ones, should be eradicated. For instance, society is unthinkable without hierarchical differentiations of conduct. Whether it occurs explicitly or implicitly, some conduct will be encouraged and facilitated, some prohibited or condemned and some enabled though not valued. Consequently, distinguishing between relations of inequality, which I treat as presumptively illegitimate, and acceptable or necessary forms of differentiation is crucial. Without it, how else will we know which forms of unequal treatment shown to particular statuses or practices should be overturned and which should not? However, framing the issue so boldly begs the question as to whether discriminatory treatment is acceptable as long as it does not maintain relational inequalities. I want to suggest that the answer is no, not necessarily; two things, here, need to be taken into account. One is the value or merit of the practice or preference in question, the other is how closely it is coupled to a form of relational inequality. But why should relational inequalities be treated differently from other forms of unequal treatment? Should not the legitimacy of maintaining or undoing them also depend on questions of worth?

This question goes to the heart of my reason for distinguishing principles of inequality from other forms of social disadvantage. While the latter are often anchored in, and work to reinforce, particular conceptions of worth, relations of inequality have played a distinctive role in shaping epistemological and normative understandings. As many commentators have remarked, how we understand, evaluate and think is largely structured and shaped, in Western societies, by gender, race and class. This does not mean that we need to find a normative position untainted by relations of inequality. Nor do we need to adopt a stance of refusing to evaluate or of only doing so oppositionally on the grounds that current values and truths cannot be trusted. However, what it does mean is that we cannot make the particular undoing of social relations of inequality conditional on judgments of value and worth. The significance of this comes into play in relation to anti-smoking policies. While harm may be a legitimate basis for introducing such policies if smoking is *not* a structural form of inequality, trusting harm and the social values that underpin it to determine the dangers of cigarette smoking is a far more dubious enterprise if these dangers and values have been generated by inequalities of smoking in the first place.

Understanding oppression

If differentiating between relations of inequality, on the one hand, and other forms of unequal treatment, on the other, is worth doing, on what basis can such distinctions be drawn? What are these relations of inequality, or oppression as they have otherwise been called? One of the most developed and influential conceptualisations of the late twentieth century is offered by Iris Marion Young (1990a). In *Justice and the Politics of Difference*, Young (1990a: 64) sets out 'five faces of oppression – exploitation, marginalization, powerlessness, cultural imperialism, and violence . . . [which] function as criteria for determining whether individuals and groups are oppressed'. Young (1990a: 64) argues that we can compare 'the ways in which a particular form of oppression appears in different groups' – that is, we can contrast the marginalisation of elderly people with that of the involuntary unemployed or particular ethnic groups. While Young's (1990a: 40) starting point is with the groups 'said . . . to be oppressed', her aim is to map oppression's different dimensions and then explore how these configure in relation to particular constituencies, such as women or gay men.

Nancy Fraser's (1997) equally influential work approaches oppression, in contrast, from the standpoint of 'social collectivities', and their relationship to particular forms of injustice. Against Young's determination to reveal the multifaceted character of the oppression most marginalised groups face, Fraser locates social collectivities along a spectrum from injustices of distribution to those of recognition. Her objectives in doing so are twofold: first, to show the tensions and differences between a politics of distribution and one of recognition, and second to suggest that, while some collectivities or 'modes of social differentiation' centre both, and are thus "bivalent" . . . differentiated as collectivities by virtue of *both* the political–economic structure *and* the cultural-valuational structure of society' (1997: 19), other modes of social differentiation – class and sexuality functioning as proximate ideal types – primarily raise injustices of distribution and recognition respectively.

Fraser's linking of particular social relations to distinct societal structures (economic and cultural), on the one hand, and forms of redress (recognition and redistribution), on the other, became the focal point for a lively debate (see Butler 1998; Young 1997b). The claims of and cleavages between Fraser, Butler and Young are also interesting in their echoing of the debates between socialist and radical feminists of the 1970s and 80s over the distinct relationship between capitalism, racism and patriarchy. In the discussion that follows, I return to some of these earlier debates. For although marginalised and discarded in the 1990s thanks to the growing influence of poststructuralism, the debates that configured around socialist feminism offer a rich seam from which to think about inequality and oppression.

While Marxist feminists saw gender as subsidiary to class (Vogel 1983: 129), and radical feminists treated sexual divisions and relations as key, socialist feminists tended to approach the relationship between economic class and gender from two different angles. The first, systems of power models, developed in the

1970s and early 80s, were largely dominated by 'dual systems' thinking in which class and gender were understood as conceptually distinct but practically intertwined (Hartmann, 1979, 1981; Walby 1990: 5–7). According to Zillah Eisenstein (1979: 22), 'capitalism and patriarchy are neither autonomous systems nor identical: they are, in their present form, mutually dependent'. Dual systems theory encountered criticism on several grounds, including for its depiction of two distinct (albeit intertwined) systems (e.g., Pollert 1996: 644–5; Vogel 1983: 129; Young 1986).[1] Nevertheless, it offered a relatively elaborate approach to the intersecting workings of capitalism and patriarchy. In other words, writers did not focus simply on subjective experiences of being female and poor, but on the distinctive, and more general, dynamics of particular systems of power. The difficulty, for systems theory, came in adding a third oppression. While black feminist analyses tended to display more ease in integrating class, race and gender, and in some cases sexuality (e.g., Combahee River Collective 1979), systems theory, which provided such a rich reading of the intersection of capitalism and patriarchy, was much less effective in theorising the interface of multiple systems (see also Acker 2000: 193).

The Combahee River Collective's (1979) early focus on a 'range of oppressions' flagged up the direction taken by much subsequent feminist work as it sought to recognise and incorporate different kinds of inequality. One major branch to emerge used the language of axes of power to conceptualise the breadth and intersections between different forms of oppression (Butler 1990: 4; Fraser 1989: 165; 1997: 179; de Lauretis 1990: 131; Sedgwick 1990: 22, 30–4). Here, not only class and gender, but sexuality and race too, became identified as distinct axes that cut across each other (Fraser 1997: 180), enabling individuals to be located at contrasting or divergent intersections. For many scholars, the advantage of the axes metaphor was that it did not place primacy on any axis *a priori* (e.g., Butler 1990: 13–14). It also allowed multiple axes to be identified. Yet, arguably this was due to its far thinner conception of the intersection between axes than that made available through a dual systems approach. In other words, keen to avoid the functionalism and restrictive conception of oppression associated with the latter framework, the axes framework ended up evacuating the theoretical terrain constituted at the point where different axes converged.

However, while theoretical work based on the axes model proved less than fully satisfactory, historical, cultural and empirical analyses were more successful, enriching and particularising the distinct interface between axes and the social. While much of this work focused on exploring how social relations condense and combine in relation to complexly situated subjectivities or identities, I want to draw attention here briefly to work which located and centred the intersections between relational inequalities in ways that extended beyond questions of identity. The field of labour and workplace relations offered one particularly fruitful area for research on complex power relations (see, e.g., Brewer 1993; Glenn 1985; Razack 1998), in part perhaps because it represented an arena in which social inequalities could be grafted on to structured occupational asymmetries (see, e.g., Acker 2000: 198).

A good example of socially nuanced work in this area is Adams's (1998, 2000) study of Canadian dentistry as a site through which social relations of class, gender and race were played out during the first two decades of the twentieth century. According to Adams (1998: 582–3), dentists of this period exploited their relatively elite class, gender and ethnic status to elevate the authority of their professional claims. Discourses of benevolent, paternalistic guidance were deployed in relation to patients who were largely white, middle-class women, and as such identified as nervous and emotional. As the remit of dentistry extended, Adams argues, discourses of degeneration and the need for cleanliness were applied in distinctively racialised ways to bring immigrants under the governance of the profession. But the impulse to extend dentists' authority did not just operate through particular discourses. Adams (1998, 2000) describes how dentists sought to raise their professional status through the employment of staff who could perform ancillary, lower status tasks. Out of this was born the lady dental assistant. The assistant was expected to demonstrate, through her self-governance, qualities that would heighten clients' association of dentistry with respectability, cleanliness and safety. The job identified as one for white, middle-class women, the dental assistant was imagined as someone who would be able to predict and respond to the dentist's needs, give attention to the multitude of minor details that distracted the dentist from his work, yet refrain from challenging or seeking to appropriate his authority (2000: 111–23).

A somewhat different response to the question of how to theorise multiple interconnected relations or structures of power can be found in the work of feminist legal scholars. Rather than focusing on the dynamic character of a field of practice, such as employment and labour relations, their work has explored the ways in which legal regulations construct, and respond to, intersecting orders of inequality. A leading scholar in this field is Kimberle Crenshaw, who has written influentially on the inability of anti-discriminatory law to respond adequately to the position of black women. Crenshaw (1989: 139) writes, 'I will centre Black women in this analysis in order to contrast the multi-dimensionality of Black women's experiences with the single-axis analysis that distorts these experiences.' Crenshaw's analysis centres on the way in which anti-discrimination law's need for a comparator, combined with its largely inadequate understanding of complex social locations, disadvantages black women in particular. Although Crenshaw explicitly focuses on the intersection of two social axes (gender and race), her analysis opens up wider possibilities in terms of studying how systems such as law flounder when social location goes beyond a single axis of inequality (see also Iyer 1993–4; cf. Eaton 1995).

Alongside the structural analyses offered by dual systems theory and its 'intersecting axes' successor lies a second approach to the question of how to conceptualise oppressions and their overlapping character; this centres on the experience of differently located groups. Iris Marion Young (1997: 17–18), one of the most influential exponents of this perspective, draws on the case of women and argues that without conceptualising them as a group, we cannot think about oppression as a systematic, structured and institutional process. At the same time, she raises concerns about attempts to map groups on to an increasingly finely tuned notion of inequality. 'This strategy can generate an infinite regress that dissolves groups into individuals' (1997: 20).[2] Deploying a 'relational' approach, Young (2000: 89) argues against treating social groups as essential in their character. 'In a relational conceptualization, what makes a group a group is less some set of attributes its members share than the relations in which they stand to others' (Young 2000: 90). Earlier, Young (1997: 27) responded to the dilemma of essentialism and the demarcation of group boundaries by developing Sartre's concept of the series as a way of identifying women's loose, shared location.

A series is a collective whose members are unified passively by the relations their actions have to material objects and practico-inert histories [practical in the sense of being the results of human action; but inert because they act as a drag and restriction on action] . . . to be part of the same series it is not necessary to identify a set of common attributes that every member has, because their membership is defined . . . by the fact that in their diverse existences and actions they are oriented around the same objects or practico-inert structures.

By focusing on the historically congealed objects, structures and practices around which series develop, Young's approach bears some similarities to my use of organising principles discussed below. However, I want to consider further here a more explicitly group-centred account of inequality and oppression in which the significance of social structure is downplayed.

Iris Marion Young's concerns about adopting a too finely tuned group-centred model have not stopped others from using group-based frameworks to explore quite particularistic social locations. In general, this kind of work has enabled a far richer understanding of subjectivity and group agency than a perspective which focuses on locations along an axis or relationally structured positions. One limitation of these latter approaches is that they tend to reduce subordinate positions to their genesis in oppression – meaningful only in relation to systems of power – and as largely passive and negative in shape. To the extent that agency of the oppressed is possible, this is understood principally in terms of resistance or social struggle. In contrast, a group-based approach, which does not reduce subjectivity to the effects of power, makes it easier to think about other forms of agency and subjectivity, including the ways in which subordinate social identities are shaped and rendered meaningful by groups themselves. Although a group-based framework highlights the ways in which subjugated positions are sustained and reproduced by the actions and practices of members – for instance, women's participation in maintaining 'demeaning' feminine norms – the generation of beneficial and productive qualities among subordinate constituencies is also recognised (Young 1993: 131–2; see generally Fenstermaker and West 2002). Indeed, pride movements oriented around race, gender, sexuality and, more controversially, class (e.g., Hartsock 1979) highlight the ways in which subordinated or oppressed identities have been consciously revalued and reclaimed, both in the pursuit of other goals, and importantly as intelligible and desirable social and cultural structures in their own right.

The limits of axes and groups

In my discussion of organising principles in the section that follows, I want to craft a framework based on the insights of both structural and group-centred accounts. However, before doing so, I want to highlight some of the problems with both of these models of social inequality. I will start with the more structural of the two accounts, focusing on the model of social axes, as this has largely replaced the earlier dual systems account.

An immediate, but resolvable, limitation of the axes approach arises from a particular reading of the metaphor. It concerns the presumption that any given social relation or axis produces only two, bipolar positions: powerful and powerless. While the possibility of being located according to a set of contradictory subject positions is recognised – as in the case of a black, heterosexual, disabled man – each position is seen as lying at one or other polarity. This obviously ignores the fact that at any given historical moment some identities may be better understood as lying *along* a race/class/gender continuum rather than at its polarity or, indeed, as unintelligible within such terms. Jewish, Irish or multi-ethnic identities, for instance, fit uneasily within a bipolar model of race (e.g., Azoulay 1997). Likewise, intersex, transsexual and transgendered people complicate a gender axis which only allows for social positioning as male or female.

The thin notion of inequality and social location proffered by the axes framework is particularly evident when it comes to explaining contradictions, such as legal provisions which appear to reverse relations of inequality (see also Smart 1992). Here I do not mean affirmative action measures which are explicitly established to counter prevailing biases. I am referring rather to provisions whose unintended consequences are to disadvantage more privileged actors. For instance, in some cases lesbian mothers have been in a stronger position than heterosexual ones in relation to welfare benefits or custody. While these reversals can be explained in terms of the occasionally beneficial consequences of remaining unrecognised and therefore outside the sweep of the state's regulatory net, they do demonstrate a quality of social relations which cannot be captured by a model that sees certain locations on an axis as invariably more powerful nor by a model which reads such paradoxes or contradictions as evidence of the axis's weakening hold.

Related to this is a second problem: namely, how do we conceptualise the place where axes intersect? As several commentators have argued, an 'additive' approach ignores the complex, often unknowable, ways in which social axes combine. At its most extreme, what may seem a subordinated identity in one context may amplify power, at least in some respects, in another: late middle-age, perhaps, when coupled with poverty or wealth respectively. An additive model, which treats each axis as distinct and separate, also fails to account adequately for the way in which individual people articulate and integrate divergent social locations. A white woman, for instance, is neither constantly located at some midway social point, nor endlessly seesawing between polarised positions of power and powerlessness. The way in which her social location is lived out is far more uneven and complex than either of these two possibilities allow (see also Razack 1999).

One reason for this conceptual difficulty in understanding the interface between axes comes from a problem more commonly associated with dual systems theory: the notion that axes have an existence apart from the ways in which they combine. But this premise of separate axes combining reveals an ontological fallacy. It arises as a result of giving analytical models the status of fact. Models that emerge as rough approximations, developed by humans in an effort to try to understand the social, become reified as phenomena with an independent and prior existence. Discrete axes of gender, class, race and age do not exist independently on some distant plane prior to their convergence in the form of distinct social permutations. Rather, identifying axes of class, gender, race and age occurs in the course of making sense of social life. Their analytical starting point, in the context of Western liberal societies, is as already entangled (see also Flax 1995). Spelman (1990:135–6) makes a similar point, arguing that oppressions are not experienced as separate packages. And, indeed, I want to emphasise that class, gender, race and age cannot be drawn out of their enmeshed state in some pure, unadulterated form.

At the same time, taken to its limit, hyper-intersectionality makes it impossible to talk about race, class, age and gender altogether since, if they are always enmeshed, how can we ever know what each of these different principles or elements contributes? In this scenario, race, class, age and gender become emptied of all meaning, rendering it also impossible to explore the effects of their combining. This is the paradox of intersectionality, yet it does not make the attempt to explore how inequalities operate and combine worthless. Rather, it requires us to be attentive to the ways in which relational inequalities can only be tentatively and hypothetically separated. I do not want to go so far as to say that, as distinct principles, class, race, age and gender are fictions – for this clearly negates their powerful sociological significance. It also ignores the fact that some social regimes or practices, at particular historical junctures, seem very largely to be accounted for, or governed, by a particular axis of inequality.

So far, the problems I have identified largely concern the way in which axes shape social location. The third problem I want to identify addresses the way in which the axes framework explains other aspects of the social. Although each axis represents a relationship of domination, once they are combined – with different individuals located at different points of intersection – the model provides no way of understanding how these different junctures relate in terms of power (or anything else). This is not just a question of how people relate across different social relationships, but of the relationship between complex social locations themselves, for instance white working-class masculinity and black middle-class femininity. By eliding people and location, the axes model fails to separate and, thus, to understand the relationship between such social locations and the humans inhabiting them. In a sense, despite seeming on the surface to offer a more structural account, the axes framework simply presents a group-based account in another name. While it offers a way of seeing group location relationally in comparison with an explanation which treats groups as social positivities, it does not really constitute such relations as socially meaningful.

The model of overlapping groups provides, I have suggested, a more nuanced conception of social identity, group formation and belonging. It also bestows upon groups an agency in structuring and shaping their relations with others, rather than seeing them as fully constituted and defined through bipolar power relations. In the chapter that follows I discuss some limitations with a group-based model from the perspective of pursuing an equality politics; here, however, I want to address two problems particular to a model which centres the subject class in thinking about inequality.

Even more than the axes approach, a group-based model finds it difficult to explain the social except as it is produced through relations between already existing classes or forces. While the axes framework sees groups as relationally constituted, the logic of a group-based framework is to place groups prior to relations of inequality. Men and women, for instance, conceptually (and historically) precede patriarchy. The limitations of this kind of approach are too well rehearsed to discuss in any detail. However, the problems of particular relevance to my argument here are twofold. First, the claim that domination or inequality is largely driven by the (false) self-interests of particular constituencies – men, or perhaps more explicitly white bourgeois men – ignores the ways in which other aspects of the social shape and influence gender relations. As a result, locating the problem in group domination generates a flawed understanding of social change, since it assumes that if the self-interest of the ruling class can be shaken or overturned, patriarchal relations will vanish. In other words, patriarchy is sustained by the interests and unequal power of gendered classes rather than by other social or structural processes.

Second, a group-first approach can produce excessively fragmented accounts which neglect more systemic connections. For instance, mapping the oppression of women and transsexuals from the premise that these are two distinct groups, albeit with overlapping histories and causal connections, can produce a narrative in which interwoven disadvantage is encountered by two separate constituencies. This neglects the extent to which the construction and positioning of male, female and transsexual subject locations emerge through common and intertwined processes. Gender may be contradictory and uneven in its workings, without a single, unifying logic, but the relationship and nexus between different gender positions (including in some cases their common ground) is not contingent or accidental. It does not arise from processes of gender working upon pre-existing, differentiated subjectivities; it arises because male, female, intersexual or transsexual are fundamentally *gendered* positions.

In suggesting that the group should not become the *a priori* starting point for understanding oppression or inequality, I do not want to imply that group identities are insignificant. I have already mentioned the socially and personally productive qualities and 'liberatory' capabilities associated with collective positions that a group-centred analysis can usefully illuminate. Less beneficially, collective identities such as black, Jewish or female, for instance, function as social facts that have been deployed and mobilised by institutional regimes to secure forms of marginalisation, exploitation or domination. Yet, even where group classifications seem most central to political regimes, we can see them as the effects of power, not its cause. This is particularly apparent in the distinctions between insider and out apparent in Nazi dicta of how much Jewish blood was required to constitute a Jew, and in the racialised classificatory processes historically applied in the United States and South Africa among others.

We need, then, an approach to social inequality which can achieve several things. First, it should illuminate the complex, contradictory ways in which multiple, elastic and socially mobile identities and subjectivities are forged.

Second, it needs to recognise the distinctive role of subordinate forces in shaping relations, identities, normative traditions and practices in ways that cannot be reduced to the reproduction of subordination. Third, it needs to find a way of exploring inequalities which treats them as something more than the domination of one group by another, and which recognises their already intersecting character. And fourth, it needs to explain how social relations such as class, gender, and age intersect other aspects of the social. In what follows, I want to propose one way of responding to these challenges. It would be rash to suggest that what I am proposing fully meets the demands made upon it. At the same time, it does provide a framework amenable to the directions and concerns so far explored.

Organising principles of inequality

The term 'organising principles' has been variously used by several authors (e.g., Alarcón 1990: 360; Flax 1987: 640; Fraser 1998: 15; Kendall and Wickham 2001; Seidman 1997: 227). In the discussion that follows, I want to elaborate on one use that builds on my earlier work (Cooper 1995a, 2002). The organising principles with which I am here concerned relate to social inequality. Conceptualising inequalities in this way highlights their 'organising' quality: principles of inequality structure, shape and pull together different aspects of social life. In other words, gender, class, sexuality, age, bodily capacity and race do more than simply name relations of subordination where 'an agent is subjected to the decisions of another' (Laclau and Mouffe 2001: 153). They also shape, and can be read off from, the allocation, deployment, effects and history of technologies of power (Cooper 1995a).

Feminist work, in particular, has explored the possibilities for reading gender relations from disciplinary and coercive technologies. Scholarship has also explored the complex ways in which institutional sites and processes, such as the state, law and family, are structured and rendered intelligible through the workings of, in particular, race, gender, class and sexuality. Organising principles of inequality shape institutional practice as the technologies of power they imbue pervade organisational life. A number of writers have explored the ways in which state bodies and agencies are gendered as a result (e.g., Naffine 1997; Newman 1995). Work has also considered the capacity of changing institutional and social forms to address and *re*configure principles of inequality (e.g., Halford 1992; Lansley et al. 1989). So, to give an example, while heterosexual norms and disciplinary power may course through local government, action at the local level can also lead new principles of sexuality to circulate through, and be read off from, municipal organisational practices (Cooper 1994; Cooper and Monro 2003). This example also highlights the competing tendencies towards divergence and isomorphism. While principles of inequality demonstrate considerable unevenness and inconsistency in their operation, there are also countervailing pressures to 'organise' inequality in ways that achieve coherence and compatibility, as I discuss later in this book.

My aim in discussing principles of inequality is to explore how they differ conceptually from other forms of social disadvantage. I therefore want to focus on three issues which help to establish their distinctiveness. The first concerns the relationship between such principles and a binary conception of power. The second relates to the way in which principles of inequality embrace aspects of the social beyond that of group membership. And the third focuses on the relationship between principles of inequality and social dynamics, such as capitalism, desire and the intimate/impersonal divide. The relationship between inequality and the pursuit of social change is dealt with in the two chapters that follow.

There is a tendency, endemic – but not restricted – to much Western feminist writing, in which relations of inequality, such as gender, class and race, are reduced to simple binaries of oppressor and oppressed. But relations of inequality do not have to take this form. Indeed, we might see the binary moment – in which oppressor and oppressed are figured as encountering each other across a *single* border – as a distinctive stage or juncture in the operation of modern, Western organising principles – presently still more solidified and entrenched in the case of gender (see Butler 1990: 7), say, than in relation to age or class. Understanding the processes by which relations of inequality become binarised requires us to examine their particular history and development, and how, prior to their emergence as binary terms, some principles may operate less as relational inequalities than as terrains, discourses or webs of practices that work to consolidate (or conversely to undermine) other inequalities (e.g., Tyner and Houston 2000). For instance, Foucault's (1979: 120–7) work on the eighteenth- and nineteenth-century European bourgeoisie suggests that sexuality functioned then, not as a relation of inequality, but as a technique of power deployed in the creation of a disciplined, differentiated, bourgeois citizenry (see also Dreyfus and Rabinow 1983: ch. 8).

To the extent that sexuality emerged in the twentieth century as an organising principle in its own right (a claim rejected by revolutionary feminists who continued to see sexuality as a configuration of disciplinary, coercive and ideological power that worked to maintain women's subordination), it has been regarded as generating two opposing classes: heterosexual and homosexual (Hennessy 2000: 100–3). Eve Kosofsky Sedgwick (1990: 2) develops this point: 'What *was* new from the turn of the [twentieth] century was the world-mapping by which every given person . . . was now considered necessarily assignable . . . to a homo- or a hetero-sexuality, a binarized identity that was full of implications . . . for even the ostensibly least sexual aspects of personal existence.' As Sedgwick and others have suggested, this binary not only proved precarious, but was itself a mobilisation of power. Thus, by the turn of the twenty-first century, this binary was itself under threat from countervailing forces, namely transgender and sex radical activists who challenged the notion that the gender nexus of desire named the main sexual division of power.

Yet, even as we begin to tell this story of a binary sustained and troubled by others, questions emerge as to whether such a binary ever really existed. The experience of sexuality in the twentieth century has always been splintered by differences of gender, as well as by class, age, and racialising processes. At the same time, while the subjectivity and lived experience of black working-class gay men and upper-class white lesbian women appears remarkably different, narratives of a sexual binary may resonate in two respects. First, they cohere with the idea of a shared – what Iris Marion Young (1997b) calls 'serial' – *outsider* relationship to the institutions of heterosexuality, even though the actual 'outsider' position may be qualitatively different. Second, the narrative of a sexual divide reflects the 'thickening' or consolidation of the hetero/homo distinction that occurred in the last century. While this was partly generated by state and other dominant forces, lesbian and gay community life and political activism – driving policy reforms in health, education, housing and social services – contributed to the deepening and extension of sexuality as a binary organising principle. On one level this can be read as the outcome of incorporation within a pastoral politics of governance which depended on lesbian and gay constituencies appearing victimised or oppressed (Cooper and Monro 2003); but it also reflected the investment and drive within lesbian and gay communities to construct a 'modern' homosexual, despite the risks and fears this simultaneously incited. Evolving and adapting as the century progressed, sexuality mobilised, while concurrently being read off from, health needs, social care, work roles, style, manners, leisure provision, erotic choices and family forms. Thus, while striated and divided by gender, class and racialising processes, the experiences of different lesbians and gay men were, if to varying degrees, 'constructed through and in relation to each other' (Sedgwick 1990: 37).

Principles of inequality influence and shape, in uneven ways, different subjectivities. They also extend further to structure, and be read off from, such things as the 'doing of gender' as it takes a cultural and institutional form (see West and Zimmerman 2002). While ideologies of gender, class and race work to normalise, rationalise and value the prevailing forms in which social relations are inhabited (as well as, to some extent, their reform), here my concern is with the ways in which norms, values and discourses associated with particular principles of inequality become attached to other institutions and processes through a process of translation or equivalence. This is a distinctive quality that principles of inequality possess. Feminists, for instance, have explored how values associated with masculinity, such as detachment, rationality and formal justice, when articulated to institutions such as law, local government and the commercial sector, help to reinforce the 'natural' affinity between masculinity or manhood and these institutional structures while also facilitating the capacity of institutions to reproduce existing inequalities and forms of dominance. Others have explored how processes such as boundary maintenance, permeability and, most obviously, penetration become gendered through the symbolic equivalence drawn with male and female bodies (see chapter 6). The capacity of norms and values associated with relational inequalities to circulate through and even to colonise other aspects of the social is not, however, restricted to gender. Seidman (1997: 93–4), for instance, writes:

Queer theorists shift their focus from an exclusive preoccupation with the oppression and liberation of the homosexual subject to an analysis of the institutional practices and discourses producing sexual knowledges and the ways they organize social life . . . by sexualizing – heterosexualizing or homosexualizing – bodies, desires, acts, identities, social relations, knowledges, culture, and social institutions.

In *Epistemology of the Closet*, Sedgwick (1990: 11) explores the way in which the homo/heterosexual divide has similarly provided 'a presiding master term of the past century, one', she argues, '. . . that has the same, primary importance for all modern Western identity and social organization . . . as do the more traditionally visible cruxes of gender, class, and race'. Without necessarily agreeing with Sedgwick on the discursive significance of sexuality compared with other principles of inequality, Sedgwick does usefully demonstrate the ways in which 'the now chronic modern crisis of homo/heterosexual definition has affected our culture through its ineffaceable marking particularly of the categories secrecy/disclosure, knowledge/ignorance, private/public, masculine/feminine, majority/minority, innocence/initiation, natural/artificial . . . utopia/apocalypse, sincerity/sentimentality, and voluntarity/addiction' (1990: 11).

The second important way in which organising principles of inequality impact on other aspects of the social is more contested. It concerns the relationship between principles of inequality and social dynamics, a term I use to refer to processes, such as capitalism, community boundary formation, desire and the intimate/impersonal, which reach across and combine social life in ways that drive or provide the motor for both social stability and change. Although, for analytical ease, I treat social dynamics as if they can be separated, this should not be read as meaning that capitalism, for instance, exists apart from the intimate/impersonal. Rather, they exist as mutually determining, deeply interconnected threads running through the social. However, the main quality that

I want to stress in this chapter is the distinctive relationship social dynamics have with principles of inequality in contrast to the far looser, less co-dependent relationship between dynamics and other forms of disadvantage or discrimination. This claim will become clearer as the discussion proceeds.

A social dynamic approach to inequality

Situating principles of inequality in relation to social dynamics is one means of avoiding the reductionism of a group-centred account which treats gender, class, race and sexuality as mobilised and sustained (but also resisted) through the competing self-interests of two antagonistic groups. It also avoids the dualism that Nancy Fraser has been criticised for constructing between economic and cultural processes, since social dynamics combine and cut across these divisions. We can see organising principles and social dynamics, as I am using the terms, as two intermeshed layers within a much more complex, never fully representable, whole. In portraying them as layers, I want to convey two things. First, principles of inequality and social dynamics form, intersect and shape each other in distinct and varying ways. Second, they can be viewed as coexisting, but alternating foci, where bringing one into sharp relief blurs the other.

The approach I am adopting thus differs from one which locates the dynamic process within inequalities of class or gender themselves (e.g., Gottfried 1998: 455). This is not to suggest that principles of inequality are rigid or fixed, but I want to create an analytical space between principles such as gender, class, race and sexuality and the social dynamics of capitalism, boundary formation and the intimate/impersonal. Changing focus can then prove analytically useful. For instance, different class positions come into view depending on whether we focus on capitalist dynamics or on the organising principles of socioeconomic class (cf. Hennessy 2000: 49). While there are clear connections and similarities between capitalism and socioeconomic class relations, they are, importantly, non-identical. Socioeconomic class relations are obviously formed by and through capitalist processes, but they are also shaped, if in more mediated ways, by other dynamics, including the intimate/impersonal and community boundary formation. As a result, principles of socioeconomic class incorporate, and are reproduced through, the doing of kinship, social capital, taste, style, and interpersonal exchanges (Bourdieu 1984; West and Fenstermaker 2002: 70–5), in ways that can be derived neither wholly from discrete group interests nor from their position within capitalist economic relations.

In making the general point that relations of inequality are anchored in social dynamics, socioeconomic class provides the most secure example, given its widely recognised (although not exclusive) relationship to capitalist dynamics. In her discussion of the relationship between class and gender, Pollert (1996) differentiates between the two on the grounds that while class is a system with its own institutional structure and motor, gender is simply a form of group dominance. In the discussion that follows, I want to take Pollert's (1996) insights on class and argue that they apply more widely: that is, all organising principles of inequality have distinct, but also multiple and overlapping, relationships to different social dynamics. I want to illustrate this point by briefly sketching one dynamic – the intimate/impersonal – that I return to through the course of the book, and which has played a central part in the reproduction of gender relations, in particular.

Feminist work of the 1970s and 1980s sought to embed relations between women and men within dynamic processes wider than just relations between groups. There was, however, considerable disagreement about the primary dynamic responsible. While some, such as Vogel (1983), focused on capitalism, other feminists located gender in the public/private divide, heterosexuality and kinship relations, with Catharine MacKinnon (1983) famously locating male/female relations in the eroticisation of dominance and submission. The approach I want to adopt, by contrast, emphasises the links between gender as an organising principle and a range of different interconnected dynamics. At the same time, I want to suggest that gender is particularly and formatively entwined with a specific social dynamic: that of the intimate/impersonal, particularly 'domestification'. It will become immediately clear, however, in describing the intimate/impersonal, that it is also inextricably entwined with other social dynamics, notably capitalism and desire (see also Acker 2000: 203).

Domestication refers to processes by which humans and animals are tamed or made at home; domestification, as I am using the term, identifies instead wider processes of home-making. This includes the relational ways in which the norms, practices, spaces and ideologies of home emerge, are constituted through, and shape the domain of the public and impersonal. It also includes the tensions and challenges that face the colonising ethos of home. The relationship between the intimate and impersonal has been extensively explored by feminist scholars from a multiplicity of perspectives, most commonly in the language of the public/private divide. Political theory, jurisprudence, economics, social policy and cultural history have all offered different paradigms through which the public/private, the gender division of labour, intimacy and domesticity can be analysed. While some work has focused on one side or other of the divide, my concern here is with the dynamic, interactive character of domestification processes. In other words, I want to move away from seeing the intimate/impersonal as a binary division and to see it instead as both a dynamic antagonism and as a complementary, intersecting set of processes. As feminist scholarship has demonstrated, this has a number of aspects. I want briefly to mention three, and then say something about their particular relationship to principles of inequality.

The first concerns the mutual constitutedness between the intimate and the impersonal: the ways in which each shapes and brings the other into being. This includes the contribution made by the domestic provision of care, moral and cultural development, food, cleanliness and security to economic profitability, public life and wider community wellbeing, as well as the ways in which government policies, broader community norms and industrial relations structure, define and contain home-life (see generally Barrett and McIntosh 1980; Himmelweit 1995). But the relationship between the intimate and impersonal extends beyond their mutual constitution to include other forms of engagement and contact. At the level of exchange, feminist work has demonstrated how women's domestic labour is exchanged for housekeeping money, while marriage itself represents a form of contracted intimacy that furthers impersonal economic relations. Contact also takes place through the way in which impersonal norms and governance techniques enter and dominate domestic life. These techniques range from legal codes and public officials to industrially generated domestic goods and services, and the effects of 'home-working' on familial relations and domestic economics.

Intrusion and permeation go both ways, as domestic skills, norms and discourses also pervade civil and political life. Practices of care provision and sexual servicing are carried out commercially, according to changing geographies and boundaries between paid and unpaid labour,[3] and according to changing perceptions of the proper relationship between the intimate and impersonal within particular professional and institutional spheres, such as healthcare and law. Less materially, the domestic has a long history of being called upon as a normative, symbolic and ideological device through which ostensibly impersonal spheres, such as the polity, nation or empire can be re-read (George 1998b: 57–8), and as a mnemonic code for selling products (Sanders 1998). Women's movements have also drawn on women's association with the domestic to validate and strengthen claims for economic and political rights and influence (see, e.g., Romero 1997), a theme explored in the contemporary, British political setting by MacKay (2001).

Domestication – the taming and civilising of subjectivities – has historically played a distinctive and deeply racialised role, in relation to white men's capacity to function effectively and appropriately within impersonal relations. At the same time, the potential colonisation of public domains by domestic norms and values has frequently been read as inappropriate or troubling. In this sense, the domestic or intimate needs to be read as dynamically striated by the anxiety endemic to its feared domination. Within societies, such as Britain, the physical realm of intimacy operates as a space to be left, at least symbolically, and at least by some, while the claims and values of intimate and domestic life need to be contained by contrasting values of rationality, the universal and impersonal. The anxiety or discomfort that besets an excess of non-domestic intimacy has a very practical manifestation. This is particularly evident in those sites where intimate or domestic conduct is seen as spilling over into inappropriate spaces, such as when sex occurs in city parks, lavatories or beaches; basic living functions are performed by homeless people on city streets; and 'inappropriate' romantic or sexualised relations (e.g., between a teacher and student) are enacted within institutions governed by non-intimate norms.

My discussion so far may sound as if I am suggesting that principles of inequality are homologous – that is similar in structure, position and role. In the course of *Challenging Diversity* I highlight several differences between them. Here, I want to draw attention to their capacity to be differentiated as a result of the particular form of their articulation to different social dynamics. This is not a simple matter. While in some geopolitical contexts a dynamic, such as the intimate/impersonal or community boundary formation, is mainly tied to a single or at most dual set of principles, in modern, heterogenous societies the position is more complex, complicating our ability to tease out the particular role of distinct principles of inequality. We can see this in the case of the intimate/impersonal. Gender here clearly plays a central role in structuring the dynamic and in filling and making sense of the roles it creates. Age too plays a part in this process, shaping not only the social positions created and the ways in which they are filled, but also the more general relationship between intimate and impersonal life.

Other principles, in contrast, such as race and class, may *seem* to work largely by structuring the allocation of gendered roles – determining which men and women can be found where. In this sense, they can be differentiated through their more mediated relationship to the intimate/impersonal as well as by their more direct, determining relationship to other social dynamics. But any suggestion of the less constitutive relationship of race and class to the intimate/impersonal also needs complicating: by the ways in which relations of race and class shape processes of domestification more generally; by their role in maintaining a heterogeneous dynamic which cannot be reduced to a smooth, unitary system; and by the historical changes which cause different principles of inequality to govern the intimate/impersonal (Comacchio 1999; Davis 1981; George 1998a; Romero1997). To the extent gender has played a leading role in structuring the roles and positions that the intimate/impersonal lays down within modern British society, social and economic changes may be causing this 'special relationship' to fade. Following this line of argument, the last two decades can be read as illustrating the difficulties caused for those gendered subjects who have become displaced as a result of other principles of inequality, particularly class and geopolitical location, coming to structure intimate/impersonal roles instead.

Nuance, complexity and an understanding of history need, therefore, to be at the fore in differentiating principles of inequality through their relationship to distinct social dynamics. At the same time, my main point is the conceptual distinction between inequalities in general, and other forms of disadvantage where this central, unmediated role in shaping specific dynamics is absent. While these forms of disadvantage may well be affected by particular social dynamics as I discuss in relation to tobacco smoking below, they do not sustain, charge, inhabit and reproduce dynamics, such as the intimate/impersonal, in the ways inequalities of gender, sexuality, race and class have done. This more mediated relationship means that disadvantages are less embedded and fortified and therefore more open to change; in particular, they do not reproduce – through the operation of particular social dynamics – their own conditions of existence.

In the section that follows, I consider the case of cigarette smoking in Canada, the United States and Britain at the end of the twentieth century. Given the crusade against tobacco that has occurred, has smoking come to constitute an organising principle of inequality? Has it become, as a result, a legitimate site for liberation? Or is it simply a mode of conduct that can reasonably be prohibited? As I suggested at the start of this chapter, identifying those forms of inequality that operate as organising principles is important. Not only does it facilitate our understanding of specific societies, since organising principles of inequality operate as core, constitutive elements, but the distinction between such principles of inequality and other forms of disadvantage is an important one for a radical, normative politics. Distinguishing between, or disadvantaging, particular conduct and social statuses may be deemed wrong on the grounds that such statuses or conduct have value or may be complicated by their relationship, however mediated, to particular organising principles of inequality. Nevertheless, discrimination or disadvantage is integral to the constitution of society, and I am therefore not arguing that all such distinctions be overturned. I am arguing, however, that organising principles of inequality should be. In part my argument is based on the claim that the distinctions such principles anchor are both mutable and unjustifiable. It is also grounded in the claim that since organising principles directly impact on our capacity to judge by shaping the values and conduct that societies privilege, we cannot trust rational processes of evaluation in determining how to respond.

I suggested at the outset that my argument for distinguishing organising principles of inequality from other forms of differentiation required a framework which set out what such principles were. I have sought to map one framework for doing this, organised around three elements: the social production of unequal positions; the ways in which principles of inequality shape, and can be read off from, other modes of power and institutions, such as the state, corporations, education and law; and the processes by which principles of inequality sustain and are reproduced by particular social dynamics.

Smokers: the new oppressed?

The purpose of anti-tobacco is the total annihilation of smoking and smokers, through any means available, including social marginalization, violence, intimidation, falsification . . .[4]

Struggles and disputes in industrially developed societies over the acceptability of smoking have been long-standing (Garner 1977). However, these struggles took on a distinctively modern shape from the second half of the twentieth century thanks to growing restrictions, particularly in North America, on smoking activity, and the rise in influence of anti-smoking stances that highlighted, among other things, risks to health (Breslow 1982; Shor et al. 1980; Troyer and Markle 1983). As a result of the growing campaign against cigarettes and the attendant commercial concerns of tobacco companies, pro-smoking lobby groups emerged, taking their place alongside other groups seeking to defend the rights of the unfashionable – recreational hunters, gun users and divorced fathers – against those perceived as 'politically correct'. Consciously and strategically, the unfashionable have argued for rights to counter what they claimed were experiences of marginality, discrimination and oppression; in this way, they sought to challenge the current distribution, as they saw it, of minority status and entitlements (Brickell 2001; Herman 1997). In the case of smoking, pro-smoking lobbies argued that smokers were not only increasingly subjected to discrimination, marginalisation and exclusion, but, within the wider polity, they were deemed *legitimately* subject to such treatment; smokers constituted the 'new apartheid'[5] – the 'new "politically correct" minority to oppress'.[6]

   To the extent that they have been able effectively to activate these discourses, smokers pose a troubling constituency for a politics that seeks to celebrate diversity. While such a politics may be fine with its traditional constituencies of the 'oppressed', equating such collective identities – implicitly or otherwise – with a broadly progressive agenda, it is placed, as I discussed in chapter 2, in a more difficult position in relation to groups whose social location is more ambivalent and whose politics appear less unequivocally worthy of support. Yet, while many writers have focused on this latter question of value, my argument in this chapter foregrounds other questions, namely of structural location.

The value, or otherwise, of smoking *is* relevant, but for the purposes of my discussion here its relevance is as a subsidiary issue to the primary question of whether smoking and smokers constitute an organising principle of inequality.

The conflict between pro- and anti-smoking lobbies has crossed several terrains, the most prominent being the discursive space of health, rights and freedom (see Avido 1986; Troyer and Markle 1983: 101; Tuggle and Holmes 1997: 85; Uyl 1986). However, the terrain most relevant to my discussion is that triggered by the pro-smoking lobby's claim that the attack on smoking represents a fundamental assault on the subjectivities, identities and status of a particular group of people. From the perspective of the cigarette lobby, an attack on smoking is also an attack on a class of people defined by their relationship to smoking in the same way that anti-sodomy laws particularly impact on gay men. According to pro-smoking author, Norman Kjono, 'One cannot be anti anything that people do without also being anti to the people who do it.'[7] Carl Stychin (1996) and others have argued that the relationship between acts and identity is a complex and mediated one – for instance, same-sex sexual activity can be engaged in by heterosexuals, and homosexuals can be celibate. However, in the main, act-based prohibition is likely to have a distinctive set of effects on actively identified constituencies. While it may lead to a closeting or covering up of the vetoed activity, it may conversely strengthen an identification – demand a 'coming out' – in relation to activities where otherwise the link between conduct and status might, as in the case of smoking, have proven relatively weak. This may be because people identify with an activity, and resist in its name, when they feel that the activity – translated as their lifestyle – is under attack (Poland 2000: 6). It is also because, like others, such as homeless people whose lives are structured by their lack of permanent accommodation (Harvey 1993: 117), bearing a smoking identity, in an anti-smoking environment, is shaped, extended and fortified by the numerous ways in which smoking impacts on how one lives, works, eats, travels and socialises. Far from being a negligible aspect of social activity, intensive smoking can create a life that diverges, cigarette lobbyists have suggested, significantly from that of the non-smoker. Initially segregated on trains, buses and aeroplanes, smoking activity has now been banished from many places. Indeed, having been turfed out of work and leisure venues, smokers are now facing removal from the open spaces in front of buildings on the grounds that they are unsightly – off-putting to customers and clients.

The process of turning people into pariahs, to the extent it has occurred, means that it is not only smokers who are increasingly identifying themselves with smoking. Anti-smokers also associate smokers with their conduct. Distaste for the visual presentation of people smoking, particularly when combined with a growing perception that cigarette smoking demonstrates a lack of self-discipline and self-respect, can generate aversion or disgust towards those who smell of tobacco, or have nicotine-stained hands and teeth (see also Tuggle and Holmes 1997: 85). This articulation is, undoubtedly, exacerbated by the links between smoking and poverty. As several commentators have argued, distaste and the intensified regulation of smoking articulate more complex social attitudes, including, in particular, the revulsion that many people feel towards the ill-disciplined poor. According to Tuggle and Holmes (1997: 90), in their analysis of a Californian conflict over a ban on smoking in public buildings, 'Ultimately, a lifestyle associated with the less educated, less affluent, lower occupational strata was stigmatized as a public health hazard and targeted for coercive reform' (see generally Berger 1986: 233; cf. Feinhandler 1986: 172). Poland (2000: 9), drawing on the work of Bourdieu, makes a similar point. 'Cigarette smoking, a mark of distinction when it was a social novelty among the well-to-do, is increasingly cast as "crass" and "vulgar" as it becomes more common . . . and therefore no longer a mark of social standing.' At the same time, despite the fact that anti-smoking policies may impact particularly harshly at both a symbolic and material level on poor people, the association of smoking with poverty has also rendered some working-class spaces less prone to anti-smoking laws. According to Poland (2000), processes of banishment have been particularly apparent in middle-class spaces, as smoking activity becomes relegated to less attractive, working-class venues, such as bars and manual workplaces.

But smokers as a constituency are not just targeted through the cultural associations of smoking and through the restrictions on where they can smoke. According to smokers' rights groups, people's subjectivities and lived experiences are directly targeted, through the discrimination, marginalisation, even demonisation, faced by a person who currently smokes, once did, or who smells as if they do. Thus, a parent's smoking may affect custody rulings,[8] while employers, in many jurisdictions, can advertise for non-smokers and lawfully discriminate in recruitment against people who smoke.[9] In Britain in the late 1990s at least two local councils attempted to require smokers to work an extra half-hour each day to compensate for the time they spent smoking.[10] Pro-smoking pressure groups also point to the inequities in health care treatment that face people with a history of smoking, and to the existence of housing providers who refuse to rent to smokers.

If we accept for the moment, then, the premise of pro-smoking lobbies that cigarette smokers are subject to significant discrimination, marginalisation and other kinds of disadvantage, would this be sufficient to identify smoking as an organising principle of inequality parallelling and intersecting other principles such as gender, class, age, (dis)ability, race and sexuality? In other words, apart from the ways in which the stigmatisation associated with cigarette smoking is both a technology and effect of organising principles such as socio-economic class, is smoking emerging as an organising principle of inequality in its own right? Pro-smoking groups, unsurprisingly, suggest that it is. They argue that people's relationship to smoking unequally affects their material experience, self-regard and individual opportunities in numerous and diffuse contexts. To this extent, it is possible to assert that smoking operates as a set of socially structured, and increasingly antagonistic, positions, albeit, I would argue, ones that are still relatively mild. However, what the lobby has ignored, or perhaps found much harder to demonstrate, is cigarette smoking's more general impact on modes of power, institutional structures and social dynamics.

Clearly, smoking at the turn of the twenty-first century does impact on some policy-making, legislation, governmental expenditure (e.g., on health and education), and personal, corporate and state revenue, yet, in general, the extent and range of cigarette smoking's impact is fairly narrow. There is no evidence that smoking, as the enactment and reproduction of socially asymmetrical positions, affects institutional forms such as education, local government, the military or the law. It also does not render modes of power intelligible, even as the regulation of smoking draws on the technologies through which modes of power, such as ideology, coercion, discipline and resource allocation, are performed (see Cooper 1995a). The distinction between technologies of power being deployed in the service of smoking regulation, and being rendered legible *through* smoking, becomes clearer if we take the case of ideology. As I suggested above, tobacco smoking carries with it a range of social meanings, but what impact, if any, does it have on the meanings and values through which other social phenomena are understood? For instance, just as we talk about institutional or national cultures being gendered in ways that reproduce the asymmetry of values and norms associated with femininity and masculinity, could we talk about them equally as being 'smoked', where the values and meanings associated with smoking are ascribed less value than their non-smoking binary counterparts?

To consider what this might look like we could start with Feinhandler's (1986) reference to purity and pollution. Other binaries might include habit/non-addiction, dependent/free-standing, odorous/fresh-smelling, weak-willed/self-controlled, buzzing/calm, cyborg/whole, chemical/natural, penetrated/closed (cf. Sedgwick 1990: 11). Clearly, some of these binaries are culturally resonant and even significant, but how dependent are they on smoking? There is certainly evidence that smoking is associated with despised and pejorative norms, evoked, for instance, in the anti-smoking lobbies' drive to outlaw 'drifting tobacco smoke' within apartment buildings, with its imagery of dangerous, cancer-causing chemicals seeping invisibly 'through a building's ventilation system and into nonsmokers' apartments'.[11] However, this imagery is not driven or significantly moulded by smoking; rather, the anti-smoking lobby has drawn on an already existing set of devalorised terms to reinforce their opposition. While it is theoretically possible that smoking could come to play a powerful part in driving cultural norms, there is little evidence that this has yet occurred.

The other way in which tobacco smoking presently fails to attain the status of an organising principle of inequality concerns its relationship to particular social dynamics. Certainly, we can link the smoking/non-smoking divide to the micro-dynamics of conduct-based stigmatisation, status-building, and the more general dynamic processes of community formation (Breslow 1982: 141). We can also highlight the ways in which capitalist processes drive both the marketing of cigarettes and the opposition shown to anti-smoking reforms and lobby groups. These social dynamics are undoubtedly significant, and play an important role generally in reinforcing existing cleavages of class, gender, race, age and sexuality. However, I want to suggest that, while they are coupled with smoking-based inequalities, the coupling is a very mediated one. In other words, smoking and the dynamics of conduct-based stigmatisation, community formation and status building can, and do, exist independently of each other. Although the links may be becoming institutionalised, particularly through public-policy-driven political geographies, they are very unevenly operationalised. Moreover, while the dynamics of differentiation, status and boundaries may structure the treatment some cigarette smokers receive, there is, again, far less evidence that the treatment meted out to smokers helps to sustain or reproduce these dynamics, more broadly, in any significant manner.

The far more mediated coupling of cigarette-smoking-based inequalities and social dynamics reduces smoking's capacity to impact on the social. One consequence of this is that the social response towards smoking and smokers is far more fluid and revisable than is the case with other social constituencies. At the same time, if we take pro-smoking lobbies' claims at face value, the relationship between tobacco smoking and social dynamics is becoming less tenuous in certain geographical localities where community formation, desire and the intimate/impersonal are being driven by the 'assault' on smoking. Let us imagine that this is case: that smoking has become more tightly coupled to specific social dynamics in ways that are mutually reinforcing. Let us add that the place smoking fills socially has become more pronounced, causing smoking-based values and norms to seep into and 'infect' modes of power and social institutions, and thereby to structure dominant epistemologies in ways that make a non-servile form of knowledge or judgment impossible. Then it may become more plausible to describe smoking, strange as it currently sounds, as having mutated into a new organising principle of inequality.

Conclusion

My aim in this chapter has been to explore a way of thinking about social inequality that moves away from the tendency to treat all forms of disadvantage as homologous, and instead differentiates on the basis of their structural character. To pursue this line of argument, I have drawn on the work of feminist writers who have offered both group-based models and axes or systems-based models. While I have outlined some of the problems with these approaches, what they together offer, in comparison with more traditional Marxist approaches, is a way of combining structural analysis with a recognition of the pluralistic character of social inequality. In this chapter, I have introduced the concept of organising principles as a way of beginning to flesh out this combination. I have also introduced the concept of social dynamic. This term is, in my view, important to recognising the depth and complexity of relational inequalities' social anchoring, ensuring that inequalities are not reduced to competing group interests. Social dynamics provide a way of examining how inequalities are fortified, strengthened and accelerated. At the same time, because inequalities are linked to multiple dynamics, with varying degrees of embrace, there is space for change, contestation and unevenness.

In this chapter I have ventured into the terrain of social theory; however, my primary reason for doing so has been a political one: to address the growing use of the list, with its anxious etceteras, to identify social divisions such as gender, race and class. Writers and activists are understandably keen to avoid practices of exclusion: the omission of a group or social axis perceived by others to be important. However, the use of the list in this way undermines our understanding of social relations.

Some lists set out explicitly to identify groups subject to exclusion or disadvantage. These lists can never be comprehensive because identities or constituencies organised around a perceived experience of disadvantage or discrimination are, for all practical purposes, infinite. And this list, depending on the premises that underpin the identification of disadvantaged groups, might include smokers too, among others. But this list of disadvantaged subjects is not the same as identifying organising principles of inequality. And I want to suggest that it is here that we need far more care and deliberation. This does not mean completely rejecting the postmodern plea that we refrain from drawing lines and distinctions that resurrect hierarchies of importance – in the process ignoring the concerns of those judged to be undeserving or deviant. However, I want to treat this plea as going to *how* we respond to different agenda. I do not want, for instance, to see anti-smoking bans which demonise smokers, which fail to tackle the links between socioeconomic class, gender and smoking, and which treat smoking as far more serious than less easily targeted forms of pollution. Identifying groups as disadvantaged but not as constituting a distinct organising principle does not mean that the disadvantage should be ignored; rather, it demands a more *evaluative* process in determining what the proper action should be. However, where organising principles of inequality, by contrast, are at stake, we cannot trust our processes of evaluation and judgment. These principles rightly sit, as I discuss in the chapter that follows, as *a priori* subjects for dismantling.

I have argued for the need to distinguish between forms of inequality that operate as social principles and those, such as smoking, that currently do not. However, I have not explicitly identified where I believe the line should be drawn. In part, I have refrained from doing so because it is a distinction that varies according to time and place. It also needs to be seen in the light of those principles of inequality that appear to be waxing or waning: in other words, over time, some principles of inequality thicken and intensify while others decline in social significance. Rather than a clear border, we might identify a grey or fuzzy space that declining inequalities cross as they move towards erasure, figuratively passing new principles moving in the reverse direction. Yet, even adopting this less hardened division may still leave this approach vulnerable to the criticism: why use as illustration an inequality, such as smoking, that seems relatively far from this grey or fuzzy border? I have suggested that smoking is, in some respects, closer than it may at first glance appear; nevertheless, there are other, arguably more proximate, examples, such as religion. I have, however, chosen smoking because my more normative and strategic concern, for reasons that will become clearer in the chapters that follow, is with organising principles of inequality in the process of coming into being. Whether smoking will ever operate in this more embedded and systemic manner is impossible to predict, although, in the dawn light of the twenty-first century, it seems doubtful. However, the ways in which it is moving in this direction (as well as the ways in which it is not) tell us something about what it takes to count as an organising principle of inequality.

Making this argument does not require us naively to accept the claims of the pro-smoking lobby, with its cynical use of political language and right-wing libertarian alliances; nevertheless, the anti-smoking movement is not innocent either. While smoking may never achieve the status of a principle of inequality, this does not take away from the importance of ensuring that unpopular activities do not produce demonised, stigmatised people. More theoretically, the progress of anti-smoking politics raises questions about the means by which new forms of inequality become embedded. When does institutionalised opposition to particular activities (inadvertently) have this effect? To what extent do social dynamics, such as community boundary formation, pick up and run with emergent inequalities? And what are the effects of embedding inequalities on other aspects of the social? As subsequent chapters explore, these concerns are vitally important to the pursuit of a challenging equality politics.

Notes

1. Vogel (1983: 170), in contrast, adopts a more unitary analysis, in which capitalism uses gender, for instance through the particular role that working-class households, and, by extension, working-class women, perform as a 'kin-based site for the reproduction of labor power'.
2. One approach which has sought to avoid this dilemma can be found in the writing of Lorde (1984), Moraga (1983) and Anzaldúa (1990). Discussing the accounts these authors have given of ways of avoiding the problem of increasingly 'narrowly defined social locations', Bickford (1997: 121) suggests, 'Refusing the split does not involve achieving a neatly unified self. It means refusing the closure of fragmentation, and recognising the specific but related "sources of living" that can be brought to bear on political action.' This perspective helps to illuminate how subjects creatively confront the reality of multidimensional identity formations.
3. Care and intimate support can also of course be provided by non-householders on a voluntary basis. One recent example of this in a number of post-industrial countries was the emergence of the 'buddy' system developed to support and care for people with AIDS.
4. FORCES, 'Understanding the parameters', http://www.forces.org/fight/files/paramet.htm.
5. Written by the campaigns director of FOREST, a pro-tobacco pressure group, *Smoking: The New Apartheid* faced criticism for suggesting that smokers could be compared with black South Africans during apartheid, see E. Mccolm, 'We're victims of apartheid say smokers', *Daily Record,* 8 Nov. 1999, http://no-smoking.org/nov99/11-09-99-4.html.
6. M. Banks, 'Thoughts on Smoking', http://w3.one.net/~banks/ sessay.htm.
7. See http://www.forces.org/writers/kjono/files/violenc3.htm.
8. http://ash.org/custody-and-smoking.html.
9. This kind of employment discrimination has been challenged by pro-tobacco pressure groups and has led in some US states to 'smokers protection laws'; these include prohibitions on employers penalising the use of lawful products in out-of-work hours. See http://www. tobacco.org/news/66362.html.
10. K. Hilpern, 'Life's a drag', *Guardian*, 19 April 1999, http://no-smoking.org/April 99/04-19-99-4.html; M. Smith, 'Smokers told to work more hours', *Daily Telegraph*, 15 Jan. 1999, http://no-smoking.org/jan99/01-15-99-7.html.
11. http://ash.org/smoking-in-condos-and-apartments.html.

4

Towards equality of power

The relationship between equality and diversity politics is a complicated one. While multiculturalism incorporates equality as a measure of minority groups' and individuals' status, poststructuralist voices operating within the terrain of diversity politics have adopted more critical perspectives. The poststructuralist emphasis on pluralism and difference precipitates fears that equality may demand or slide into sameness, and, as such, prove antithetical to freedom. Consequently, one primary aim of this chapter is to think about equality in a way that strengthens and consolidates its political purchase, while taking seriously these concerns, namely that equality neither jettison freedom nor require sameness (see also Flax 1992; Phillips 1999). My strategy for doing so is threefold: to take the individual as equality's 'who', power as equality's 'what', and relations of inequality as the domain of equality's 'how'.

Politically, equality advocates have largely been divided between those arguing for a liberal approach based on ignoring 'irrelevant' characteristics so that otherwise similarly placed individuals can be treated alike, and proponents of a group model. While some advocates of the latter have linked equality or justice to the irreducible differences between groups, others have simply treated the group as the meaningful subject in treating like alike (see Ward 1997). Although I remain unconvinced that equality requires sameness in some or other respect (cf. Ward 1997), my starting point is the moral equality of living humans. In the first part of this chapter, I differentiate this approach from liberal individualism as well as from group-centred accounts. The second part goes on to address the question: equality of what? Much recent thinking in this field has revolved around the question of distribution and recognition. While both approaches have merit, I want to suggest an alternative framework based on equality of power. This approach also has limitations. However, the aim of my discussion is not perfectionism, but to use equality's 'what' in a somewhat different way. This becomes clearer in the third part of the chapter, which focuses on the question of how. My argument is for the pursuit of equality of power through the undoing of inequalities of class, race, gender and sexuality. However, such an undoing, when it confronts the very real existence of gendered, racialised and sexualised forms of difference, becomes highly contested. Among proponents of diversity politics, views differ as to whether such an undoing is possible, as well as whether it would be beneficial. In the final part of the chapter, I seek to concretise this discussion by exploring the different constructions of what counts as gender (in)equality between feminist and transgender activists; in the process, I ask whether, at the level of strategy, there can be any common ground.

Towards an equality of selves

Within liberal societies today, the dominant model of equality is that of formal equality, understood as the elimination of arbitrary and unjustifiable forms of discrimination between individuals. This model of equality has been extensively debated, with its anchor in the cognitively coherent subject whose qualities, including the capacity to exercise freedom (Cornell 1998: 18), render her, but usually him, worthy of equal respect – at least from the perspective of formal governmental action (see generally Dworkin 1985: 190; 1994: 273). Yet, even to the extent that liberalism's formulation of equal respect provides a common currency between people, foregrounded in equality's emphasis on shared characteristics, dignity and societal membership, liberal equality is also anchored in a second, more antagonistic trajectory: that of the competitive individual. From this perspective, not only do subjects see their own wellbeing as both independent of and, more fundamentally, as threatened by the wellbeing of others, but equality itself constitutes a finely tuned balancing act between subjects.

Diversity politics approaches equality from a different perspective. Given the complexity and breadth of this political space, it is unsurprising that it gives rise to a range of tensions, including between equality and social pluralism, and equality and individual freedom. At the same time, the emphasis placed by diversity politics on collective identities colours its understanding of equality. This has taken two primary forms. The first suggests that since social groups play a significant role in creating and maintaining inequality, they are a key site for social action in the pursuit and attainment of equality. The second states that the group is central to equality's pursuit because of its importance in shaping people's wants, beliefs, commitments and identity. In other words, since people's lives are lived, and made meaningful, through communities, equality is both unintelligible and unachievable unless seen in intra-group terms.

Both these claims, which arise from the discursive structure of diversity politics, are important in their insistence that we cannot think about equality outside the social processes, forces and claims that structure our lives. In our interpersonal relations, few can separate their interests, needs and desires at the 'dermal' boundary.[1] Nevertheless, I want to propose a framework which, while placing collective structures, and in particular organising principles of inequality, at the centre in thinking about equality's 'what' and 'how', decentres the collective in relation to equality's 'who' to focus instead on the equality of persons.[2] My reasons for doing so are threefold. First, I want to avoid the elisions and erasures that occur when a larger class is identified as equality's subject – where the treatment of parts become less important than the whole, so that equality between communities does little to address inequality within. This has been identified as a serious problem in relation to socially heterogenous communities, where leaders who reflect the interests of dominant forces within the community articulate the group's needs and wishes. My second reason stems from the fact that communities or social constituencies are not discrete, unconnected entities. Given the growing recognition of intersection, of the ways in which people are located at the juncture of several social relations, group equality is unable to capture and respond to these complex locations. It can therefore lead to the problems discussed by Crenshaw (1989) where individuals, seeking legal remedies to discrimination suffered, become obliged to identify convincingly the discrete group inequality at stake (see also Iyer 1993–4).

At the same time individuals are not reducible to their social identity. This is my third reason for adopting an individual-based conception of equality, rather than a group one. Irreducibility is not simply explained by processes of intersectionality or by the infinite surplus that exceeds any particular identity description. It is not simply that we cannot account for every aspect of a person's identity – that if we could do so, subjects could be fully known and described; rather, I want to make the more fundamental point that social location is not fully constitutive of being. This does not require us to resurrect a universal subject, nor is it based on the existence of a social gap either now or in the past enabling us to point to those parts of living that are the product of a residual human beingness and those that comprise and are forged out of identity. What it does mean, rather, is that we can imagine a figurative space – or perhaps an elasticity – in the ways in which people inhabit, craft, struggle over and reflect on the 'who'ness and 'what'ness of being. Creating a space between identity and being is important to the process of imagining the possibility of identity changing. If there is nothing beyond identity, as some feminist writing implies, if we are all and only constituted as the particularistic subjects of (intersecting) social relations, political challenges to particular identities can only be interpreted as an attack on being. History demonstrates the way in which attacks on particular identities have often invoked – as either a preferred course of conduct or a last resort – the erasure of being. However, I want to suggest that being and identity can and need to be distinguished. Judith Butler (1990: 147) makes a similar point,

> the reconceptualization of identity as an *effect*, that is, as *produced* or *generated*, opens up possibilities of 'agency' that are insidiously foreclosed by positions that take identity categories as foundational and fixed. For an identity to be an effect means that it is neither fatally determined nor fully artificial and arbitrary.

Butler develops this point in the context of different kinds of gender identity performances and productions. However, I want to use it to make a somewhat different claim. Namely, if equality focuses on individuals rather than social constituencies, it allows us to explore what individual equality demands of social identities and relations. This is not a politics of playing God – evaluating and juggling the character and existence of social constituencies from on high. Rather, it calls for a far messier, 'in the dirt' politics, one willing to engage with social location and identity in a way that takes account of – indeed is driven by – the inequalities of power and psychic investments at stake, but does not allow these to delimit social change by treating current identities and constituencies as a fixed given.

If the subject of equality is the human being, what, then, is equality's object or content? This question lies at the heart of discussions of equality which have extensively debated different answers to equality's 'what'. I do not want to re-trace these footsteps through equality of opportunity, rights, freedom, outcome, satisfaction, but instead want to focus on two other responses to the equality question. Equality of resources and recognition have become particularly prominent in turn of the century debates in Britain and North America, covering as they appear to do the terrains of economics and culture respectively (cf. Yar 2001). In the previous chapter I cited the debate triggered by Nancy Fraser's work on inequality and injustice. Here, I want critically to consider resources and recognition as frameworks of equality. Using Ronald Dworkin as a proponent of the first, I draw on Nancy Fraser's work in my discussion of the second. Fraser's (2001) later work clarifies the space between her status-based approach to recognition and more conventional group-based approaches. My discussion focuses, however, on the latter in order to highlight the difference between equality of recognition and equality of power, discussed below.

Equal resources and recognition

Equal distribution of resources offers one of the most common interpretations of equality, encompassing questions of both distribution and redistribution. As Nancy Fraser (1997: 25) argues, equality of resources can take a reformist or moderate form through the systematic repetition of 'surface reallocations';

alternatively it can involve a more radical transformation in the way in which resources are produced and acquired. While the first approach, with its second level transferrals – or compensations – from the wealthy to the poor, tends to embed and legitimise prevailing economic dynamics, such as capitalism, posing little apparent threat to socioeconomic class locations, the second seeks to overturn dominant socioeconomic relations by challenging the economic frameworks that anchor them.

To explore the model of resource equality further, I want to consider the work of the influential legal philosopher, Ronald Dworkin. Dworkin may seem a curious choice given the fact that his work on resource equality is driven by a far less radical agenda than that of other authors discussed in this book. However, in his extensive and elaborate exploration of equality, Ronald Dworkin's work usefully illustrates the problem with resource-based models that sit on top of existing socioeconomic structures, namely their failure to engage adequately with the social. This is apparent in Dworkin's work in the course of his theory-construction, the distinction he draws between tastes and disabilities, and in the lack of regard paid to non-resource-based drivers of social inequality.

According to Dworkin (2000), resource equality constitutes a system of allocations where nobody would envy anyone else's resource package (including their labour) taking a lifetime as a whole. This approach allows for diversity driven by different tastes and choices, while emphasising the importance of treating people as equals. In *Sovereign Virtue*, Dworkin (2000: 4) asserts that arguments about equality should begin 'in our life and experience', and he operationalises this assertion in the latter 'inside-out' part of the book. Nevertheless, Dworkin replicates the tendency of many resource-based equality models to treat equality as a kind of formula or paradigm that can be derived, at least in general terms, through a combination of philosophy and economic theory. Rather than taking the social as the starting point for what equality means and what its pursuit entails, Dworkin, like others, draws on hypothetical scenarios: shipwrecked survivors washed ashore on a desert island, for example.[3] It is out of these 'primitive' societies or hypothetical narratives that basic principles are derived (and subsequently complicated), and then installed within more complex liberal societies. This does not mean that the social is entirely absent. Alongside its more covert presence in shaping the models generated, for instance Dworkin's notion of the auction, the social also provides the dilemmas that form the meat of equality theorising. But what, above all, is missing in all of this is an overt account or theorising of the social. How we think about the social makes a crucial difference to our understanding of present-day asymmetries, visions of alternatives, and the plausibility of different equality strategies; yet this material foundation is missing in many more philosophical accounts.

The asocial character of Dworkin's resource equality model stretches beyond the methodological absence of social theory. It also lies in the way in which choice is framed. Dworkin is concerned that his framework should distinguish between expensive tastes and unchosen disabilities. His reason for doing so is to avoid a situation in which people with 'champagne' tastes are given extra resources to allow them to have the same level of satisfaction as someone with what we might call 'plonk' tastes (Dworkin 2000: 55–8). At the same time, Dworkin is anxious to ensure that a model which does not compensate for expensive tastes simultaneously does not disadvantage people whose costly needs derive from a 'handicap' (Dworkin 1995: 277, 294). While the reasons driving Dworkin to establish this distinction are understandable, his framing of it ignores the role of the social in the construction and evolution of needs and preferences. So, expensive tastes are distinguished from disabilities, not because one or other is socially constituted (or because both are in different ways), but because the former is chosen and the latter is not. The erasure of the social is thus also evident in Dworkin's naturalistic understanding of disability. According to Dworkin (2000: 297), 'having a physical or mental infirmity or condition that makes pain or depression or discomfort inescapable . . . is . . . an evident and straightforward handicap. Someone with such an infirmity did not choose it; he [sic] would cure it if he could.'

Dworkin's use of the term 'handicap' asserts an unproblematic norm against which people are measured and compensated (or treated differently) when they fall 'below' through no fault of their own. Thus, 'special needs' are depicted as non-social facts that should be dealt with, where possible, through remuneration, not by changing (or diversifying) the normative assumptions of embodiment that anchor and underpin the social. Adopting a more social understanding does not mean denying the physical reality of being unable to walk unaided or of experiencing visual difficulties. At the same time, as disability activists and scholars have argued, it is the norms and infrastructure of a society which disadvantage people who diverge from their morphological assumptions (see, e.g., Fawcett 1996; Freund 2001).

Equality of resources models, such as Dworkin's, see 'disabilities' as lacks that can be treated by the bestowal of additional support because of their inability to see social structures. In other words, equality can be achieved through redistribution with its implicit assumption that inequality is created by the unfair allocation of a fungible, transferable and ubiquitously beneficial substance. While resources can have a wider meaning, they do not seem to in Dworkin's work, apparent in his claim (2000: 70) that the market is a crucial way of ensuring that resource equality works, while 'the device of the auction might provide . . . a standard for judging how far an actual distribution, however it has been achieved, approaches equality of resources at any particular time'

(Dworkin, 2000: 72).[4] Auctions and markets take preferences as given, and assume that resources are exchangeable. As a result, they do little to tackle the ways in which inequality is produced and sustained through the factors driving taste and preferences, on the one hand, and the role played by disciplinary and coercive forms of power, on the other.

[I]t is unjust that some individuals and groups are denied the status of full partners in social interaction simply as a consequence of institutionalized patterns of interpretation and evaluation in whose construction they have not equally participated and that disparage their distinctive characteristics or the distinctive characteristics assigned to them. (Fraser 1998: 24)

While resource equality largely focuses on the individual, the paramount concern of much writing on equal recognition is the group or collectivity. Nevertheless, it can be related back to individuals to the extent their equality depends upon how groups of which they are members are seen and treated. According to Nancy Fraser (1997: 14), injustices of recognition are 'rooted in social patterns of representation, interpretation, and communication. Examples include cultural domination . . . ; nonrecognition . . . ; and disrespect'. In her discussion of cultural imperialism, Iris Young (1990a: 59) makes a similar claim, 'Those living under cultural imperialism find themselves defined from the outside, positioned, placed, by a network of dominant meanings they experience as arising from elsewhere, from those with whom they do not identify and who do not identify with them'.

Emphasis on recognition and representation poses a challenge to the more conventional stress by the left on resource distribution. Equal recognition distances itself from, on the one hand, the homogenising models of right and left in which difference equals deviation and, on the other, more structuralist accounts where relations of domination appear so all-consuming that subordinate constituencies consist of little else. The politics of recognition, in contrast, is grounded in the centrality of group membership to individual esteem and identity, and in the importance and capacity of disadvantaged constituencies to experience a positive collective sense of self (Taylor 1992; cf. Fraser 2001). In other words, while relations of domination and oppression have structured the ways in which subordinate groups are seen, they have not diminished the capacity of such groups to possess value.

The politics of regard and recognition have been described as driving a new political hegemony that has at times overshadowed distribution-based concerns (see, e.g., Fraser 1997; cf. Yar 2001). Yet, in considering the limitations of equal recognition, I do not want to assume that advocates treat it as the sole equality or justice principle. Young (1990a), for instance, presents cultural imperialism as one of *five* faces of oppression, and few would argue that recognition is the only thing that matters. Nevertheless, even as a fragment of a wider politics, recognition as a normative framework is riddled with tensions, particularly evident in its attempt to straddle a pluralist conception of the social and a more universalist emphasis on shared norms and inclusivity. Fraser (1998: 33) frames this tension in terms of what is being misrecognised: 'In cases where misrecognition involves denying the common humanity of some participants, the remedy is universalist recognition. Where, in contrast, misrecognition involves denying some participants' distinctiveness, the remedy *could* be recognition of difference.' However, I want to suggest that this tension is not easily resolved. Aside from difficulties in distinguishing between 'common humanity' and 'participants' distinctiveness', tensions between a universal and more particularistic or pluralist approach permeate the issue of who recognises and on what basis. These tensions are present even if recognition is reframed in terms of institutional rather than individual or group action (see, e.g., Fraser 2001).

Commonly, recognition depends on the values of the one who recognises. How the other is seen cannot help but be mediated by the habitus of the one who does the looking. But to what extent can the regard or recognition they proffer be 'borrowed' from those to whom it is given? I raised this issue earlier in relation to the eruv, where proponents argued that non-users did not need to understand the eruv; they simply needed to respect the fact it was of value to orthodox Jews (see chapter 2). This kind of approach became more pervasive within feminist politics of the 1980s and 1990s in opposition to what were seen as universalist feminist claims (e.g., Okin 1994, 1999; cf. Flax 1995). Yet, while there are good reasons for arguing that disempowered constituencies should set the terms on which they are recognised, a pervasive problem with this approach is its tendency to assume a common, shared position within the constituency or group itself. In addition, it inclines towards the reverse of what recognition intends by privatising or 'walling off' accountability. As a consequence, it risks substituting formal permission for 'real' recognition.

An alternative response to the question of ethical regard or recognition decentres the question of who it is that is establishing the terms on which recognition takes place to focus instead on the underlying norms and values through which such recognition is constituted (cf. Fraser 1998: 24–5). Galeotti (2002: 104–5), for instance, suggests that public recognition is anchored in accepting the possibility of different viable options within a society without implying that any particular difference is valuable or beautiful. But what counts as viable and worth recognising depends on values other than those of affirming difference. This becomes apparent in relation to the awkward minorities for diversity politics: recreational hunters, conservative religious members and smokers. What kind of recognition should meaningfully be bestowed upon them? Should it depend on judgments of worth, or, as Fraser (1998: 37) suggests, on the impact of their actions and participation within a particular context? And how, given the emphasis placed upon *public* status within equality of recognition, can the values of those with less power gain effective 'airtime' in shaping what counts as 'equal respect', 'participatory parity' (Fraser 2001) or the opportunities for 'achieving social esteem' (Fraser 1998: 36)? For the danger is that these criteria become saturated by established and dominant notions of value.

Equal recognition raises questions about who it is who recognises, as well as on what basis, and according to what values, recognition takes place. Linked to these two questions, however, is also a third: *what* exactly is being recognised? If sexual orientation, for example, is always lived out across, and in relationship to, other social principles, how can it be 'recognised' in isolation? How do we identify the gay, black or female aspect of a person's identity?

There are two approaches we might take to this. The first focuses on the different ways in which social relations are lived in all their complexity. The trouble with this strategy, however, is that it can fragment into a claim for recognising each person's unique combination of identities. The second response is to abstract a (minority) sexuality, race or gender from lived experience, and construct a set of generalised or shared characteristics, practices and history that can be valorised. The obvious problem with this latter approach is that it may end up privileging a particular version of what it is to be female, black or gay (see also Appiah 1994). It also raises difficulties in relation to the question: equal to what? Is being black supposed to be equal to being gay or female? Or is the equality in question anchored in the validation dominant subjectivities receive, so that being black should have equal recognition to being white? But if we recognise the radically different locations of black and white subjectivities within histories of colonisation, exploitation, discrimination and privilege, on what basis can regard be equal? Here, I am not concerned with the impossibility of overturning existing inequalities, but rather with the difficulty of determining what in fact dominant locations should receive recognition for. On what basis, for instance, should whiteness receive regard?[5]

Dominant identities are in a paradoxical position, invisible in many respects while receiving an excess of positive regard in others. A politics of equal recognition might trouble this situation in two ways, first, by adopting a more critical kind of regard towards dominant subjectivities – raising their visibility as privileged and powerful in contexts where their taken-for-grantedness renders them unmarked. Second, to the extent that all publicity is good, equal recognition might seek to decrease the visibility of dominant forces and identities in other respects. This kind of 'negative' equality politics has received little discursive attention within mainstream multicultural movements, although more radical approaches such as revolutionary feminism, intent on the deconstruction and practical undoing of hegemonic identities, adopted versions of this stance. However, for many reasons, not least their posing of an unapologetic counter-common sense, negative recognition has not translated into effective action. More commonly, practical attempts to equalise 'downwards' adopt a more prosaic form: reducing the presence of dominant bodies from political positions, educative materials, media or cultural texts. Yet, while it is relatively easy to count and remove certain bodies and faces (despite the opposition this kind of strategy can generate when sites of power appear threatened, such as electoral shortlists), adopting a stance which disregards, neglects or reverses the cultural tropes of white, middle-class masculinity is far harder to achieve.

In raising problems with recognition as a foundational equality claim, I do not want to dismiss it altogether. Recognition speaks to, and resonates with, the concerns of important social movements. For while we may want to disclaim the pure categories of gender, class or race that a politics of recognition often avows, social movements work by naming, creating and solidifying clusters of identity. These identity clusters may be precarious and tentative, always conditional and ready to implode, but they provide a provisionally agreed basis for distinctiveness and recognition – albeit a recognition which sits less easily with those on the margins of the identity category. Yet, one dimension lost when the emphasis is placed upon group recognition is the process of making the organising principles of power themselves more culturally visible. While an exclusive focus on relations of power can prove disempowering for those subordinately positioned, treating identities as if they exist largely in isolation from their social conditions of production offers a very attenuated form of recognition. By failing to grasp how and why particular identities have come into being, it offers a thin choice between simple acceptance on the basis of what *is* or judging and evaluating others as if their artefacts, values and lifestyles are discrete, unembedded acts of self-creation.

Equality of power

Both equality of resources and equality of recognition have something to contribute to a progressive or radical politics. However, they offer, at most, a partial foundation. I therefore want to pose a third possibility, equality of power, which offers, in my view, a more comprehensive model than resource and recognition based frameworks. Equality of power gets at the socially constitutive character of inequality in ways that a resource-based model fails to achieve, and unlike recognition frameworks – which can lose a sense of who is being offered what – it holds on to why equality is important. At the same time, equality of power is striated by ambiguities and tensions. It is also, I would suggest, ultimately unattainable. But the improbability of witnessing equality of power in practice does not undermine its value as an ethical or political guide which articulates the normative premise that nobody has an inherent right to impact more on their social and physical environment than anyone else. This premise embraces the liberal and pluralist emphasis on individuals' right to an equal capacity to achieve desired ends or act in particular ways. It also includes a more radical emphasis on equalising participation within the making and operationalising of collective decisions – political, economic, environmental and social.

Before discussing equality of power in more detail, let me say something about the conception of power I want to use. It is based on a Foucauldian framework, but with some revisions reflecting my somewhat different focus (see Foucault 1980: 119; also Cooper 1995: ch. 2). Foucault provides a useful starting point for two primary reasons: first, he emphasises the structuring capacity of power, rather than focusing solely on 'power over' or power as prohibition. Second, he makes room for people's complex relationship to power in contrast to models which divide people into those who exercise power and those who have it exercised upon them. For Foucault (1983: 220), power is something that flows through the body politic, evidenced in the way in which people's agency is structured by the (institutionally embedded) actions of others (see also Simons 1985: 81–2).

Power must be analysed as something which circulates, or rather as something which only functions in the form of a chain. It is never localised here or there, never in anybody's hands, never appropriated as a commodity or piece of wealth. Power is employed and exercised through a net-like organisation. And not only do individuals circulate between its threads; they are always in the position of simultaneously undergoing and exercising this power...
(Foucault 1980: 98)

Foucault's rejection of the idea that power can be stored, and his view that its exercise is inherently asymmetrical, sits uneasily with the idea of 'equalising' power,[6] an issue I return to below. However, before doing so I want to signal three ways in which my approach diverges. To begin with, I am more concerned with the one who exercises power than many Foucauldian accounts, which tend to focus on the relationship between technologies of power and the one subjected to it. I am also concerned not just with the ability to impact on others' actions but with the exercise of power in relation to the physical environment, institutional structures, discourse and ideology. While Foucauldian studies have analysed these factors, they tend to be viewed as techniques of power; in other words, their primary relevance is as ways of affecting the actions of others, rather than as politically important objects of power's exercise in their own right.

Second, contra Foucault, I include the exercise of force within my understanding of power (see also Cooper 1995). According to Foucault (1983: 220), 'A relationship of violence acts upon a body or upon things; it forces, it bends, it breaks on the wheel, it destroys, or it closes the door on all possibilities.' In contrast, a power relationship requires that '"the other" (the one over whom power is exercised) be thoroughly recognized and maintained to the very end as a person who acts. . . . [S]lavery is not a power relationship when man [sic] is in chains.' (1983: 220, 221). Although Foucault's approach to power emphasises its relational qualities, within his framework the capacity of the one subjected is crucial to determining whether power has been exercised. However, this capacity becomes less crucial if the focus shifts away from the relational character of power's exercise to evaluating the relative capacity of different people to exercise power (including the capacity of some but not all people to coerce in a given situation). From this perspective, the ability both to shut off another's agency and to engineer effects through their unwilling bodies (e.g., forcing a woman to have a caesarian operation to protect the foetus) becomes a vitally important form of impact.

My third distinction relates to the relationship between capacity and the actual exercise of power. If, on the one hand, capacity becomes collapsed into power's exercise, it follows that decisions *not* to act become unreadable except in terms of power's lack. This risks over-interpreting decisions to refrain in a way that disallows for the possibility of reasonable, legitimate choice. It also makes a politics of equality far harder to contemplate, since it has to work against all the conjunctural factors that shape whether or not power is exercised in any given circumstance. However, collapsing exercise into capacity, conversely, carries other dilemmas. For instance, what aspects of capacity can be equalised (Foucault 1983: 217)? While some writers have been keen to distinguish 'natural' talents – omitted from the imperative to equalise – from unfair social advantages that should be so eliminated, they tend to limit the latter to resources and opportunities unjustly arrived at. This assumes that the value of certain abilities and the asymmetry of their distribution is relatively fixed. It also tends to ignore the more subtle or taken-for-granted ways in which principles of inequality structure people's capacity. Determining whether capacity is equal when the exercise of power remains asymmetrical is immensely hard (see also Phillips 2003). Liberal scholarship that emphasises the importance of formal equality between men and women, for instance, has often ignored the cultural, social and disciplinary factors that not only shape capacity, but also shape the conversion of capacity into action.

Capacity and exercise need to be retained as distinct but interrelated issues. At the same time, while social disparities in how different people translate capacity into action function as an important site of political concern, the answer does not always lie in increasing power's exercise among those who frequently refrain. Rather, as some indigenous communities, feminists, peace activists and environmentalists have argued, the gap between capacity and exercise, the choice of whether and how to act, offer important opportunities for reflection.

Such reflection and dialogue might usefully generate a decline, revisioning or degree of restraint among those who tend most often to convert capacity into exercise.

I want now to turn to some of the difficulties equality of power raises, but before doing so, let me offer a brief summary of its claims. Equality of power is predicated upon the normative premise that political and social action should engender greater equality in people's capacity to shape the social and physical world, whether as consumers, decision-makers, workers, mobile bodies, house-holders, producers, crafters, interrogators or trouble-makers. Currently, this is denied, not just by the inequalities of skills or talents on which liberals such as Dworkin place so much analytic store, but more importantly by the way organising principles such as gender, age, class, race – and, at the international level, geopolitical location – skew people's ability to exercise power. It is also denied by the institutional structures of law, government, public provision and commerce, which similarly amplify the differential impact people have. In some respects, my conception of equality of power resembles the notion of positive freedom, in that it centres the social and institutional support required for people to participate politically, economically, culturally and socially, recognising that this participation may be contingent on maintaining (but also developing and updating) collective resources and practices such as minority languages. By focusing on capacity, equality of power does not assume similitude in the actual productive effects any person will generate or wish to create; equality is not coterminous with sameness at the level of identity, practice or the good life.

Exploring the ambiguities and tensions that surface in thinking about equality of power is instructive in relation to the twofold task of developing a framework and to flagging up the political role it might play. I want to focus here on two kinds of difficulties (see also Cooper 1995a). The first are shared by other equality frameworks and result from the process of linking equality to an identifiable criterion, whether it be power, opportunities, resources, welfare or recognition. The second set of problems relates directly to the fact that it is power being equalised.

One of the main challenges levelled at equality is that it requires us to be able to quantify and measure that which is being equalised. How, though, can we determine whether people's exercise of power is similar either in terms of quantum or degree? Dworkin (2000), in his model of resource equality, tries to address this through his conception of envy. Applied here, we might say that equality means not envying the power to which someone else has access. This allows for variation in the kinds of power different people have; it also avoids the problem of the detached or external quantification of power, since it allows each person to come to their own view of when the balance has been achieved. However, the problem with subjective accounts is that they cannot easily deal with the fact that the power people wish to exercise will itself be shaped by the social conditions in which they find themselves. Thus, to the extent these are unequal, a subjective account is likely to amplify those inequalities already in existence.

The dilemma of equalisation, and the envy method, can also be approached from a different perspective that addresses the timeframe across which equality occurs. Should power simply be equalised between individuals taking their lifespan as a whole, so that no individual would envy or wish to substitute another's lifespan? Or should power be relatively equal at any given juncture – a 'society-centred' approach? The first perspective allows for considerable social inequality, provided that individuals move between different locations over their life as a whole. At its most extreme, it would permit slavery, so long as master and slave switched around at some hypothetical 'half-time' (see also Kappel 1997). The alternative, society-centred approach, by contrast, decentres the question of each individual's power quantum to focus instead on the kind of society generated. In other words, it works from the normative position that a society with greater equality of power within any given sphere, as well as at any given temporal moment, is a good in itself, and is not to be entirely judged by whether it generates quantifiable equality between individuals. My own position is closer to this latter approach, although its focus on social structures and processes suggests a less subjective and less morally pluralist standard than Dworkin's envy method. However, one area in which a lifetime approach captures something of the nature of equality is in relation to children. Children tend not to be treated as entitled to equality of power vis-à-vis adults because their immaturity renders them liable to making 'poor' judgments, particularly in terms of jeopardising their future wellbeing. While immaturity cannot be relied on to evaporate over time, it is most Western children's temporary status as such that makes their lesser power more acceptable. In other words, the broad assumption is that over a lifetime everyone can expect to experience the full spectrum of age-based fluctuations in power.

The second cluster of issues and difficulties that I want to raise relate directly to the idea that it is power that is being equalised. I have already alluded to the concern that equality of power seems to imply that generating effects is beneficial. Underlying this concern is also a more fundamental proposition: that equality of power misses the point – what matters is not how much impact anyone has, but the *kind* of impact. Indeed, critics might show the absurdity of equalising power by pointing to accidental or even injurious forms of exercise. Is the capacity to generate these kinds of effects embraced by the project of equalising power?

Accidental or unintended consequences are heuristically useful to thinking about equality of power. On one level, unintentional consequences may clearly appear to be the effects of an actor's exercise of power. At the same time, they have little to do with the motivation for pursuing greater equality of power which relates to the conviction that people have an equal moral entitlement to participate in *shaping* their environment and world. Arguably, this aspect is clearer if equality of governance – with its connotations of guidance, purpose and intentionality – is substituted for equality of power. Yet, while governance readjusts the balance towards questions of motive and agenda and away from an exclusive focus on means and effects, it leaves hanging the complex relationship between the two. Foucault highlights this point in a personal communication to Dreyfus and Rabinow (1993: 187) when he suggests, '"People know what they do; they frequently know why they do what they do; but what they don't know is what what they do does"'.

The second related issue of intentionally harmful outcomes also goes to the core of what equality of power means and entails. I suggested, at the start of this section, that equality of power incorporates the presumption that people have an equal entitlement to pursue their preferences. On its face, then, equality of power seems to suggest that those persons promoting 'good ecology', for instance, should be placed on a par with polluters in terms of their respective rights to impact on the social. This reading of equality of power clearly turns it into a nonsense; it also assumes two things: first, that equality in this form provides a discrete and trumping value; and second, that equality of power means formal parity.

One way of dealing with the problem of harmful effects is to argue that since harm reduces another's capacity to exercise equal power, such effects can be legitimately restricted. This is a popular approach. However, there is a danger of generating an infinite regress as each act protected from injury is seen as the cause of another's restricted capacity. For instance, the fascist stopped from speaking on the grounds that her utterances restrict the capacity of others identifies both the obstruction and others' prior speech acts as in turn impeding her speech. It may therefore be more helpful to adopt an approach which reads equality of power through, and in relation to, other social values, such as care, respect and responsibility – values, in certain circumstances, which will trump claims to individual parity.

At the same time, a formalistic interpretation of equality, that is, one which reduces equality to treating like alike, can be avoided – at least at the level of political strategy – by tackling inequality as a set of institutionally structured social relations. From this perspective, one environmental practice is not the same as any other, in the same way that multicultural and racist speech acts cannot be grouped together because they all involve verbal or textual utterances. Equality of power is not a present-day reality; consequently the strategies demanded need to engage with social relations as they are. Sherene Razack

(1999: 34) makes a similar point when she writes, "The individual . . . is constituted by a set of power relations cast like a net over how she or he sees and thinks. It is precisely these power relations that are obscured when we balance competing claims."

To sum up this part of the discussion, I want to propose two simultaneous moves. The first is to contemplate equality of power as an unattainable, but nonetheless necessary, aspiration, whose contradictions and tensions offer a productive surface for debate. One of the errors, in my view, of certain strains of anti-perfectionist poststructuralism was taking the claim that power and inequality would never end as a reason for binning ideals judged to be unrealisable. My second move is to suggest that equality of power can be pursued, in a way that also tackles some of the difficulties raised above, through a more structurally engaged response. This response does not seek to transcend the social through equality frameworks which reduce 'real life' to a low-level problem of implementation; rather it seeks to embed itself within the tangle and interconnections of particular social moments in order to identify, target and undo the organising principles of inequality that currently operate.

Undoing gender

If equality of power provides the justificatory grounds for challenging socially embedded inequalities, what does this challenging require? Does it entail the wholesale dismantling of organising frameworks of gender, race and class, for example, or can, and should, certain aspects be salvaged? Can gender, race and class remain as meaningful, but non-pernicious, forms of difference? It is generally accepted that socioeconomic class cannot take an egalitarian form, and, among socialists, this has tended to lead to demands for wholesale economic reform (or, at the very least, more substantial resource redistributions). But race, gender and sexuality are often portrayed as detachable from the inequalities they currently encode; in other words, relations of domination, exploitation and marginalisation are not seen as inherent to the differences they express (e.g., Phillips 1999: 26). While certain injustices, such as economic class, require category elimination, others, such as sexuality, are perceived differently (Fraser 1998: 13). In other words, gays, lesbians and heterosexuals are irreducible to, and capable of outlasting, heterosexist norms and relations.

A similar assumption lies behind the commonly held position that equality between men and women is possible. In other words, gender can – indeed should – outlive its relations of production. For some, this position is predicated in a separation of the biological from the social; while the latter is historically contingent and as such amenable to change, this does not undo the biological difference upon which relations such as gender are predicated. For instance, Anne Phillips

(1999: 26) writes, 'It is clearly inappropriate to make sexual equality depend on sex change operations that convert all the men to women or all the women to men'. Others reject notions of biological immutability, but put forward other, more normative reasons for holding on to non-hierarchical gender difference (e.g., Flax 1992: 194). In the case of gender, it is suggested, the withering away of all social distinctions between men and women would be undesirable. Gender differences are pleasurable, exciting and sexy, androgyny tends to be modelled on masculinity (Flax 1992: 196), and the elimination of diversity more generally would lead to a stagnant, flat society. Against these claims, others argue that a progressive androgyny (or 'gynandry') does not need to resemble masculinity, particularly if it is part of a wider feminist project which values 'feminised' qualities such as listening, caring and responsiveness (see also Ferguson 1991). There is also no reason why the undoing of gender and other presently asymmetric differences should necessarily produce a boring, homogenous society. While it is, unfortunately, likely that new principles of inequality will emerge to take the place of those in decline, other more benign forms of difference might also arise, including personal eccentricities, legible in ways not reducible to structural social relations.

In the final chapter of this book, I return to the question of difference in conditions of equality. What I want to consider here is the ways in which organising principles of inequality are framed and the effect that this has on political strategy that seeks to undo their significance. The example upon which I draw is that of gender, and in particular the divergence between feminists who seek to dismantle the hierarchy between men and women, and transgender activists who seek to dismantle the hierarchy which privileges the expression of fixed dimorphic genders over more fluid and multiple genders (see generally Jeffreys 1990; Monro 2001; cf. Raymond 1982). This latter, less familiar challenge grew out of a transgender politics which sought to go beyond passing and being recognised within a newly assigned or chosen gender. Instead, radical transgender or 'trans' activists and scholars sought to signal the oppression that faced people who migrated without passing or who sought, for political or personal reasons, to occupy gender in ways that could not be read uncompromisingly as either male or female (see generally Califia 1997; Namaste 1996; Stone 1991; Whittle 1996).

In considering these competing notions of what gender as an organising principle of inequality entails, I also want to explore some potential common ground. The need for alliances around a shared position has been strenuously advocated by sex radicals and queer feminists, such as Pat Califia (1997: 241, 242; see also Kaveney 1999). However, the framing of this common ground has tended to be based upon a discourse of shared experience and positioning as social marginals and deviants. It has thus both valorised and embraced sex workers, fetishists, sado-masochists, transvestites, transsexuals and 'queer' homosexuals against those seeking respectability and inclusion within the status quo. Yet this celebration of transgression and deviance has its own problems. I am therefore more interested in the strategic possibilities of an agenda intent on countering both forms of gender hierarchy; I leave hanging for now the question of what kind of gender might outlive the dismantling of the hierarchy between men and women, and the divide which naturalises fixed dimorphic genders while denying the plausability and equality of fluid, polymorphic alternatives.

I want to take from feminism its understanding of the social power and embeddedness of gender, as well as its critique of masculine norms. At the same time, I take from trans and queer politics the role played by individuals and more 'bottom-up' community-based exchanges and practices in contributing to the ways gender is lived and understood (e.g., Pratt 1995; Wilchins 1997). This form of crafting is perceived within trans, cultural feminist and queer communities as empowering and productive, and resonates with the trans claim that gender, divorced from biological destiny, can be understood as a series of grammars, modalities or communities which make (new) legible choices meaningful and possible.

The first strategy is that of 'gender morphing'. This should be distinguished from, on the one hand, drag when it functions as a purely staged performance (see also Namaste 2000: ch. 1), and on the other, from trans as an attempt to pass. While trans writers such as Rubin (1999) are concerned that others collapse trans and drag into each other in ways that ignore the specific sense of embodiment and 'realness' felt by transgendered and transsexual people, both drag and passing share ground to the extent that each can be read as practices that build and depend upon a prior or original gender. They also, albeit in quite different ways, both rely upon and actualise a commitment to a dimorphic gender system. This contrasts with the deeper refusal to be comfortably gender legible expressed by more radical trans activists.

Many feminists have expressed scepticism about the value and viability of individual gender morphing; however, I want to suggest that a radical trans politics that critiques gender as a fixed dimorphic structure and that pursues invention (e.g., Pratt 1995:184; Stone 1991: 296; Whittle 1998) can make a significant contribution to undoing gender (see also Butler 1990, 1993). It does so in two primary ways: first, by challenging gender's foundations in a dimorphic structure, and second, by affirming the centrality of choice and voluntary gender identity. But if morphing works through the construction of illegible genders, what actually counts as incoherence in this context, and how do we know when it has occurred? Does it depend on contradictions among superficial indicators: name, clothing, explicit self-portrayal, for example, or is it about more subtle or structurally embedded gender characteristics? And if the latter is the case, do not many of us engage in gender morphing when we reject or fail to conform to stereotypical or expected behaviour?

Without in any way diminishing the value of the gender non-conformist behaviour that many, if not most, feminists, for example, demonstrate, I want to argue that there is a difference which lies in how behaviour and subjectivity is read. Given changes in the way conduct and appearance are interpreted, it is possible to be a butch lesbian, for example, and still be read as a woman. What gender morphing and incoherence suggest is that, according to the norms and rules that prevail at a given time and place, individuals are not intelligible according to a binary gender structure. Because this structure not only subjugates other genders but denies them visibility and plausibility, the mere existence of other gender combinations can be seen as disrupting this monopoly in ways not achieved by gender performances which are able to be accommodated within the existing dimorphic structure. It also challenges the acute marginalisation faced by those who do not subscribe to its laws.

Yet if morphing is to have much effect, beyond revealing a minority of people whose gender identities are not easily legible, it needs to mobilise far more people around discourses and practices of gender choice. However, even among those engaged in body and appearance morphing, gender does not function as a free choice. All sorts of factors shape people's gender identities, and the idea people could be persuaded to go against that which feels 'internally' right – whether it is to morph or not to morph – as part of a political project is unlikely to prove much of a success. But generating choice is not the only issue at stake. Of more concern is whether transgendering works to recuperate and legitimate gender by redefining it as a product of choice if not of birth. From this perspective, transgendering functions as the postmodern retrieval of gender distinction from the waste-basket to which biological, involuntary differences are being increasingly discarded. At the same time, it does seem doubtful that complexly gendered lives can rescue modern gender divisions, particularly those relating to inequalities of power. The emphasis on choice and fluidity, to become another gender or avoid being legible at all, however limited it really is in practice, undermines the capacity of gender to structure relations of oppression and exploitation, since these largely depend upon immutability and a relatively rigid dimorphic socialisation.

The second strategy for undoing gender inequalities that addresses the dichotomies critically identified by both feminists and trans activists takes a different approach. Rather than seeking to disrupt and replay everyday social conduct, it targets the ways in which governments and employers interpellate and discipline citizens through gender-based classifications. Unsurprisingly, given the practical implications of denial, achieving the right to be officially classified according to one's new corporeal or attributed gender has proved an important struggle for transgender and transsexual people. Sharpe's (1997, 1999) trenchant critique of the ideologies deployed by courts to determine gender gives additional weight to the argument that the elimination of classifications might prove a more satisfactory strategy than liberalising the right of 'transition'. In relation to birth certificates, passports, driving licences and ID cards, the marking of gender according to a dimorphic grid frequently seems to serve little valid purpose, while the pro-transsexual campaign for gender transitioning to be officially confirmed, necessary in a context where classifications carry power, alienates many feminists who see such a move as consolidating gender categories. Minnie Bruce Pratt (1995: 162) vocalises this tension in relation to her partner's attempt to get a driving licence which identified her as a man, 'you needed . . . a piece of plastic with the pronoun that matched your looks so a policeman on a lonely road at night would not say, "You – get out of the car"'.

Advocating a strategy based on eliminating gender classifications has to take into account several things. First, it does not mean getting rid of all official uses of gender, for doing so might work to obscure levels of discrimination. In certain situations, for instance, in relation to pay, job recruitment, political participation or service provision, a demographics of gender can be important, highlighting potential problems and the need for reform. Lack of such information may undercut a commitment to change, as policy-makers in the field of lesbian and gay equality are finding where ambivalence about asking people their sexual orientation means that a lack of statistics exists. However, the categorical information needed is not always clear-cut. In relation to lesbian and gay policy-making, where people experience abuse or discrimination often because of another person's judgment about their sexuality, actual orientation may be less relevant than how it is perceived. Likewise, in the case of transgendered people, their status as transgendered may prove more relevant to uncovering discrimination and harassment than whether they have an entitlement to pigeonhole themselves according to their 'new' gender (see also Namaste 2000: 41–2).[7]

Classification has also been legitimised by progressive forces on the grounds that it enables diversity to be recognised, that without it people are assumed to share the needs and traits of dominant groups. A similar argument is made in the context of positive action, that without the knowledge that people are members of subordinate constituencies, affirmative action cannot work. However, like the detailed statistical information that tends to underpin and justify it, classification is of little value in general; the benefits to be drawn from placing people in categories, as well as the categories themselves, depend on the particular issues at stake. Ethnic minority forms and positive action around race, for instance, often erase ethnic identities perceived not to suffer economic disadvantage. Likewise, it is not clear whether it is necessary to 'know' that someone is a woman before pregnancy-related leave can be granted, although knowing that someone is pregnant may have clear policy relevance in certain contexts.

A strategy which simply seeks to remove classifications may challenge the subordination neither of women nor of polymorphically gendered persons. At the same time, the use of official classifications is a potential site of convergence for feminist and trans politics, given the ways in which classifications police people's gender identities and enable discrimination to occur. Rather than such a site being one of tension between those who seek to have new gender identities recognised and those who question the possibility of authentic genders, we might locate some common ground in the proposed elimination of official gender designations except for those cases where classifications are intended to redress existing inequalities.

Conclusion

In this chapter I have argued for the importance of equality as a social virtue. I began by arguing for an approach which takes the individual, rather than the group, as equality's subject. I then considered two versions of what equality's object might entail, those of resources and of recognition. While the first fails by being too divorced from the social, and, as a result, ignoring many of the forms of power the social generates, the second seems to dissolve at the point where we ask: what quality or aspect of recognition is actually to be equalised? I therefore went on to consider a third approach: equality of power. More comprehensive than the other two, it also has to address hard questions about equality as well as about power.

One response to this predicament, which some poststructuralist radicals have adopted, is to reject equality. Another is to say that equality of power is simply the wrong equality. A third is to suggest that we need more than one variant of equality to do the work. The approach I have taken, however, is to propose equality of power between individuals as a broad, but unrealisable, ideal, while focusing more specifically on how we might undo the social organising principles that currently work to make people's capacity to exercise power unequal. However, my reasons for advocating the dismantling of principles of inequality is not just in the pursuit of individual equality of power. I also work from the normative premise that the good society is one not based on structures of domination and exploitation, even if in comparing one lifetime with another we were to find more individual *in*equality than if we adopt the implausible notion of switching places at 'half time'.

But while it is an easy thing to advocate the undoing of social inequalities, understanding what this demands is often less straightforward. My brief discussion of the relationship between trans and radical feminist politics was intended to highlight some of the tensions and cleavages that can occur when the dichotomies and asymmetries that particular organising principles, such as gender, encode are contested. In this case, feminism's emphasis on the subordination of women to men confronted the emphasis of transgender activists and scholars on the dominance of a dimorphic gendered structure and the erasure and subjugation of those inhabiting gender as a polymorphic fluid set of possibilities. What follows from this is not only disagreement about the character of struggles around gender, but also the possibilities available for reimagining gender. While some feminists have argued that equality of power requires the dismantling of gender altogether, transgender activists have tended to stress instead the importance of retaining gender as a cultural and aesthetic structure through which choices, preferences and relationships can be rendered meaningful.

Yet, despite these differences of analysis and political aspiration, my argument in this chapter has been to suggest that common ground can be found. A central tenet of my analysis is for radical equality politics to move away from its conventional centring of the group, with its dynamics of inclusion/exclusion and boundary maintenance, to focus instead on norms and values on the one hand, and social relations on the other. From this perspective, radical feminist and trans politics might find some shared terrain, given their mutual interest in destabilising gender norms and attenuating the distinctions of power that gender maps. Gender morphing and declassification provide two possible tactics in this wider struggle – tactics that challenge the discursive consolidation of gender binaries and inequalities, in particular. These tactics are by no means more important than tactics that directly tackle gender discrimination or seek to reorganise social dynamics, such as the intimate/impersonal, with which the binary of male/female is particularly tightly fused (see chapter 3). They also do not address the values that a move away from gender classification and immutability might signify. However, I chose these two tactics because they speak to the anti-disciplinary ethos and concerns of a postmodern transgender politics, while also having the potential to be inflected by, and articulated to, a more normative feminist agenda.

In the chapter that follows, I explore the relationship between equality reforms and social norms in more detail. The focus of my discussion is the gathering momentum internationally for lesbian and gay spousal recognition. In exploring the politics of lesbian and gay social movement demands and the reforms that have been generated, I ask what contribution these developments make to equality politics. While my concern in this chapter has been with the pursuit of individual equality of power through the dismantling of organising principles of inequality, my focus in chapter 5 is on the relationship between equality reforms and social norms. Drawing on normative principles of proper place and the public/private, I explore the impact they can have on equality's pursuit. To what extent do they undermine the struggle for equality of constituencies, such as lesbians and gay men? Do they provide a mechanism that allows inclusion and greater power for some at other constituencies' expense? And to what extent are normative principles themselves reinflected through law reform campaigns and initiatives, such as the introduction of gay marriage and domestic partnerships?

Notes

1. Thanks to David Delaney for this term.
2. I do not therefore apply equality directly to institutional, corporate and voluntary organisations. How a multinational, local co-operative, school or temple are treated raises crucial equality concerns: for instance, does public policy protect and enhance the profitability of large corporations, while doing little to aid not-for-profit organisations or local co-operatives? But this concern relates to the way in which the treatment of different organisations and corporations impacts upon equality between individuals. Artificial subjects, such as firms and voluntary sector bodies, do not make meaningful subjects, from the point of view of equality, in their own right.
3. Resource equality models do not transcend their societal origins. Despite the potentially radical implications of determining resource allocations according to the equal respect or value of persons, liberal resource equality models seem to generate models of the good society strikingly similar to the society from which they originate – a somewhat surprising outcome, perhaps, when theorisation begins with a desert island.
4. Dworkin (2000: 80) does suggest that physical powers are resources, since they are used to make something valuable out of one's life. However, because they cannot be manipulated or transferred to create physical or mental equality, they do not constitute resources – as far as Dworkin is concerned – for a theory of equality in the way that material resources do.
5. One way out of this dilemma is to adopt Fraser's (1997) more radical deconstructive approach towards recognition. But if collective identities are pulled apart, and pulled to pieces, is it useful to refer to this as recognition?
6. Although his conception of power as something both exercised and undergone does imply a kind of balance or symmetry.
7. As Namaste (2000: 42–3) suggests, 'catch all' categories may be unhelpful where they blur relevant distinctions. What these are will depend on the particular context and issue, but may include whether someone is male-to-female as opposed to female-to-male, as well as the varying ways in which transgendered people self-identify.

5

Normative encounters: the politics of same-sex spousal equality

> The logic of marriage . . . is little more than a rationalization of privilege and will contribute to greater, not less, inequality within the lesbian and gay communities, as well as in the wider society. (Carrington 1999: 223)

From the mid-1990s, partnership recognition and gay marriage emerged to dominate lesbian and gay politics around the globe (see, e.g., Wintemute and Andenæs 2001). For many gay civil rights activists, spousal status had become a fundamental equality demand; some even declared it to be a basic human right (e.g., Wolfson 1996: 82). At the same time, others adopted a questioning attitude towards the clamour for official recognition (e.g., Boyd 1999; Herman 1990). For many critics, marriage remained a symbol of and means of perpetuating gender oppression, privatisation and state control.

In this chapter, I explore spousal recognition from the perspective of equality. However, in doing so, I want to move away from the question of whether marriage or spousal rights benefit lesbians and gay men as a class. For reasons explored in the previous two chapters, I find a class or group-based paradigm of equality politics, with its emphasis on raising a defined constituency to the standard experienced by dominant groups, unhelpful. At the same time, I want to avoid an approach which evaluates marriage and spousal recognition as if they exist in a vacuum. What spousal rights, and the effects of their expansion, mean depend on the way in which legal reform maps onto and intersects with other social processes. This more socially contextualised approach is important for understanding the effects of reforms on inequalities of sexuality. But it is also important for understanding the effects of reform on other inequalities as well. The premise underlying this chapter is that reform cannot simply be evaluated in terms of its impact on a single social inequality. Following my discussion of equality politics in the previous chapter, the effects of reform on the spectrum of inequalities that exist are as important.

For the most part, writing on equality, from a diversity perspective, has tended to ignore equality's relationship to other social norms. By this I do not mean a failure to balance. Hierarchically ordering values, placing them on scales to determine which should prevail or carry the most weight, is frequently undertaken. But what I do mean is that far less attention has been paid to the way in which values, embedded and materialised as social norms, strengthen or undercut equality's pursuit. In this chapter I argue that incorporating an analysis of social norms into discussions of equality is important for three reasons. First, dismantling particular forms of inequality often demands that the norms that anchor, legitimate or otherwise sustain the inequality be challenged. For instance, it may be hard to undo inequalities of age without paying a critical regard to norms of appropriateness, entitlement and normality. Second, norms often function as bridges between different inequalities. Aesthetics of appearance, for instance, work to consolidate inequalities of gender, race and age, as well as sexuality. Consequently, tackling one form of inequality by trying to recast its normative assumptions, for instance emphasising the attractiveness of young, butch white women rather than seeking to problematise – or radically rewrite – norms of attractiveness, may entrench other existing inequalities or ease new ones into existence.

Third, the approach described in chapter 4, with its articulation of equality to socially undoing, rather than restructuring, principles of gender, race, class, age and sexuality may seem rather nihilistic. While the effects of undoing existing inequalities may be hard to know, a politics which seeks *only* to dismantle has its limitations. Laclau and Mouffe (2001: 189) make a similar point when they state, 'no hegemonic project can be based exclusively on a democratic logic', that is, extending 'the egalitarian imaginary to ever more extensive social relations' (Laclau and Mouffe 2001: 188), 'but must also consist of a set of proposals for the positive organization of the social'. What these should be cannot be determined from equality alone – even if equality is identified as a norm rather than a strategy of undoing. As Nagel (1998: 12), for instance, argues, 'Equality can be combined with greater or lesser scope for privacy, lesser or greater invasion of personal space by the public domain'. Articulating, or rather *re*articulating, equality's relationship to other norms is a necessary way of injecting some substance into what equality means – to flesh out a richer conception of the 'better', if not 'good', society.

To explore the relationship between social norms and equality in more depth, I want to develop the framework of organising principles established in chapter 3 to consider what I am calling normative or 'is-ought' principles. Like the principles of class, gender, race, age and sexuality discussed earlier in the book, these normative principles also organise, can be read off and lie condensed within social structures. However, while the former revolve around an inequality of power between subjects, normative principles, or rather *dominant* normative principles, are defined by their capacity to condense the space between description and social vision.[1] In other words, how society is constitutes, with minor revisions, how it should be. Dominant normative principles are both definitional of a society, and a means of consolidating and anchoring it. They not only collapse – or read – 'ought' from 'is', but also seek to present a deeper, more fundamental and authentic truth about the present and good society.

In my exploration of the way in which normative principles both frame, and are framed by, same-sex spousal recognition, I focus on two principles that surface through the course of this book: proper place and the public/private. These two sets of principles are deeply relevant to inequalities of sexuality; they also have a wider resonance in relation to asymmetries of race, gender, class and age. Proper place and the public/private are firmly embedded in liberal conceptions of the social, but they are also open-textured. While they work largely to anchor and rationalise dominant social processes, they also offer more progressive possibilities. This comes from their capacity to be substantially revisioned; however, it also emerges as a result of a more immanent form of critical practice: bringing society closer to the best or truest interpretation such norms can bear.

I begin by setting out in more detail how normative organising principles work, and then turn to offer a conceptualisation of proper place and the public/private. The second half of the chapter considers the encounters between these norms and same-sex spousal recognition. In exploring the effects of this encounter, I am concerned not only with what happens to the pursuit of lesbian and gay equality, but also with how this encounter impacts on other social relations (gender, class, race, age), as well as on norms of proper place and the public/private themselves. I argue that the extent of any mutual accommodation, status quo maintenance or troubling depends – at least in part – on how spousal recognition is argued for, operationalised and inhabited.

Normatively organising the social

Normative principles do not exist separately from the social; also they do not have a life as discrete principles. As with organising principles of inequality and social dynamics, any disentangling is purely heuristic and intended to aid analysis. Principles such as democracy, liberty and fairness can be read off from the social; at the same time, their material and discursive presence works to organise and reproduce the social in distinctive ways. The notion of 'normative' in this context highlights several things. First, it identifies those principles that signal the good society. Although principles may vary by social sphere – indeed such a division between spheres is itself a formative normative principle – in the context of liberal society, principles such as democracy, rationality and accountability operate as idealisations of current social reality. Second, the good society is perceived as one that continually strives to improve on the realisation of its normative principles. Current liberal society thus provides the foundations for an idealisation which is then turned back on the social present as a guide, measure and aspiration. Third, normative principles guide individual and institutional actors who strive to conduct themselves and their society well. They are principles to safeguard at the level of the social, and to pursue, recognise and promote in everyday activity. Finally, the notion of normative underscores accounts of the principles themselves. Thus, principles such as justice or property have value because of the ways in which they are underpinned by other right-sounding principles.

Yet, principles such as normality, legality and the proper do not simply function as guides or ways of establishing the worth of conduct or the good society. Part of their power or force comes from their epistemological quality as norms or standards, as well as from the way in which the epistemological is articulated to the normative.[2] As epistemological principles, they are seen as offering not only a framework for exploring how particular societies (seek to) work, but also as something deeper, in that they identify a fundamental truth about the 'good' self and society. Nagel (1998: 10), in his discussion of the importance of privacy *qua* selective intimacy, reveals the normative power that comes from such truths when he states, 'No one but a maniac will express absolutely everything to anyone.' More generally, what this suggests is that structures such as democracy, property and accountability are not just facts embedded within a contingent present, but are necessary, constituent elements of 'good' society. With 'is' and 'ought' closely coupled in liberal society's conception of itself, normative principles play a key role in explaining liberal social formations: we cannot 'know' liberal societies without understanding the core principles that constitute them (conceptually and historically). While many such principles are recognised as contested or subject to readings that emphasise unevenness or contradiction, the struggle to provide the best or fullest interpretation represents a major project for many liberal scholars (see also Dworkin 1986).[3]

Normative principles function as key rhetorical and strategic elements within everyday political discourse. At the same time, it is important to recognise that 'is–ought' principles have more than simply discursive applicability; they do not need to be uttered to have effect. Normative principles are embedded in and realised through the preferences, desires, tastes and choices they shape (not always predictably), as well as in institutional structures such as parliamentary, legal and economic processes where democracy, accountability, property, legitimacy, fairness, justice and liberty constantly circulate and are enacted. A crucially important quality of 'is–ought' principles is that they are co-referential. Their power or force comes from the way in which they flag up, or are validated by, each other. Thus, legitimacy may be anchored in account-ability, property in fairness and liberty, fairness in justice, and so on.

We are so used to the application of these principles that their productivity is rarely apparent. However, the capacity of normative principles to facilitate and ease, indeed to generate, social processes and actions in the first place are immediately evident in their counterfactual, where conduct is rendered legible in non-normative terms such as coercion, arbitrariness, inefficiency, selfishness and impropriety. Action which presents itself, or more usually is read by others, in these ways pokes out awkwardly, provoking hostility for seeming to challenge the quiescent normative basis for conduct. Recognising the existence of these less glowing motivations and incitements to action is central to a perspective that refuses to see the social simply in the terms of its own self-image. A society may speak about its origins, ongoing practices and aspirations in normative co-referential terms, but this does not mean that these are the principles that *actually* govern or underlie institutional structures or practices (see also Cooper and Monro 2003). This is the claim of the social critic who sees normative principles as means of gaining consent or acquiescence, of mobilising people in the name of qualities whose status is illusory.

Yet despite the validity of this scepticism, normative principles are important because they provide the tools by which a society presents, evaluates and at some level, at least discursively, organises itself. Before going on to discuss my two principles, I want to draw attention to three aspects of the general char-acter of normative organising principles. The first concerns their unevenness, inconsistency and capacity to be recreated. In particular, I am interested in the capacity of normative principles to be inverted so that their aspirational dimen-sion functions as both a critique of the present and an incitement towards societal change. Because such counter-normative principles are far less naturalised than their more hegemonic alternatives, they tend to exist on the plane of political and social utterance rather than practice. However, this is not exclusively the case and, as I discuss in chapter 8, in particular, counter-normative principles can be found embedded in alternative institutional and social structures. I re-fer to these alternatives as 'prefigurative', to highlight the ways in which the norms articulated are read through a vision of reform as well as a critique of the present. In other words, they fundamentally invert the elision drawn between 'is' and 'ought'. The second aspect concerns the relationship between different kinds of organising principles. I suggested above that normative principles cite, refer to and support each other; at the same time they do not always cohere. Principles collide, colonise and threaten, raising a host of questions about how the tensions between them should be, and are, resolved (see Laclau and Mouffe 2001: 165–6). As I discuss in this book, the relationship between diversity and equality illustrates one form this collision can take. While my argument is that diversity and equality can be articulated in ways that render them compatible, for instance by reading diversity through the lens of equality (see chapter 9), other radical readings stress the tension between them.

The third aspect of their character concerns the complicated relationship between normative principles, social dynamics and social inequalities. The ways in which normative principles and social dynamics intersect, and the effects they have on each other, are important to thinking about their stability as well as the capacity to undo and vary social inequalities. One instance that illustrates this relationship is the example of the intimate/impersonal and normative principles of proper place and responsibility. As I discuss in chapter 3, the intimate/impersonal has historically been associated with producing and giving effect to gender inequalities. These inequalities have, in turn, been reinforced and stabilised by norms of proper place and responsibility which work to legitimise and rationalise the gendered allocation of roles, relations and places. At the same time, such normative principles have also been drawn on in the drive to change men and women's relationship to intimacy, domestic work and paid employment, by, for instance, reframing what men's responsibility for their children entails.

In the discussion that follows, I draw on these three aspects of the way in which normative principles work: their unevenness, contradictory quality, and capacity to change and be inverted; their relationship to each other; and the way in which they intersect social dynamics such as the intimate/impersonal in the course of exploring the pursuit of spousal equality for lesbians and gay men. But first I want to set out in more detail the two normative principles which I will discuss, namely, proper place and the public/private.

Revisiting proper place and the public/private

Proper place – with its cultural and social division between that which is in and out of place – highlights the ways in which inequality is secured through forms of differentiation and segregation that are read as civilised, natural or otherwise beneficial (Cresswell 1996). Within modern Western societies, we can see proper place (alongside other normative principles) as working to separate activities and peoples into hierarchically related, albeit mutually determining, spaces (e.g., Razack 1998). Indeed, the proper operates relationally to define and constitute place itself. In Western, liberal societies, it is a deeply and thoroughly internalised structuring device, epitomised in the emphasis placed on children knowing what goes and belongs where. Politically, the power of the 'proper' as a normative principle is threefold. First, it works to delegitimise certain distributions or combinations of persons, practices, spaces and identities – a process whose effects I explore in more detail in the following chapter in relation to nuisance. Second, it offers a powerful device for resisting change. Configurations of the proper, in which the proper is articulated to norms such as property and legality, organise, rationalise and justify dominant social practices. Third, in the way in which it underpins and is read off from physical zoning which keeps phenomena apart, the proper defuses and contains challenges.

The notion of *spatial* differentiation as a significant normative principle at the turn of the twenty-first century may seem, to some degree, counterintuitive. Despite the intensification of migratory controls and scrutiny, national spaces appear, in many ways, more culturally diverse and heterogenous; from a gender perspective, men and women seem less confined to separate spheres, and lesbian and gay sexual expressions seem more visible than even two decades ago. Yet these trends do not negate the countervailing drive for spaces to become more ordered, efficient and mono-functional. This does not require us to contrast the present with a golden age of spatial heterogeneity, but rather to attend to current impulses to segregate and discipline particular acts, movement and identities (Edensor 1999). In line with a policy rhetoric of equal opportunities, this impulse tends to focus less on status or on those characteristics discursively constituted as immutable, such as gender and race (although age remains firmly subject to injunctions of propriety). Nevertheless, the alternative policy emphasis on 'voluntarily chosen' conduct, presentations and lifestyles maps onto distinct socially identified constituencies. This is apparent in the treatment of poor and homeless people in countries such as the United States and Britain, as I discuss in chapter 6. Policies to outlaw and move on the growing numbers of vulnerable people from gentrifying city spaces have precipitated the development of technologies of ownership, surveillance and design – street benches that cannot be lain on, for instance – that render some people's very bodies improper. Likewise, norms of proper place, allied to discourses of 'visitor' and 'host', have been deployed in Britain to discipline and domesticate the conduct of minority ethnic constituencies, immigrants and refugees (see Cooper 1998a: 63).

For lesbians and gay men, the focus of this chapter, ideologies and practices of proper place have been particularly apparent in the spatial and temporal zoning of sexual identities and activities. At its most overt, this has meant banishing during the day certain activities, interactions and identities from city streets, penalising the enactment of a lesbian or gay identity at school or in the family home. While the boundaries of propriety and appropriate conduct can be explicit, with clear penalties or punishment if breached, boundaries may also function more covertly (Valentine 1996: 154).

Yet the governance of sexuality does not work only through the social dynamics of boundary maintenance, and lesbian and gay identities are not simply despatialised. While earlier scholarship explored the extent to which the proper domain for homosexuality, post-decriminalisation, was the private sphere in which scarcely tolerated intimacies could exist – if not thrive – more recent writing, reflecting the changing spatialisation of sexual identities, has explored the wider embedding of lesbians and gay men's proper place. While much of this writing has focused on the development of lesbian and, particularly, gay urban spaces and residential areas (Bouthillette 1997; Brown 1995, 2000: ch. 3; Davis 1995; Grube 1997; Myslik 1996; Quilley 1997; Rothenberg 1995), other work has explored the rural expression of gay and lesbian sexual identities (Kramer 1995; Phillips, Watt and Shuttleton 2000).

The documentation and analysis of lesbian and gay spaces does not invariably lead them to be identified more widely as proper. However, there has been a shift from associating gay sexual identities with residualised spaces to seeing the marking of space as gay as instrumental in the construction of cosmopolitan localities (Florida 2002; cf. Moran and Skeggs 2004). Thus, gay becomes a signifier and instrument in achieving normative, gentrified urban space. This development has also had its critics, particularly those who read the changes as co-opting monied, mobile, out gay men at the expense of other gay and lesbian constituencies (see Bell and Binnie 2003). For the latter, the organising principles of proper place continue to have, and be read through, more residualising effects – shunting less desirable people, activities and places into marginal, less visible and less prestigious spaces.

If proper place currently legitimates and naturalises, while also being read off from, the allocation of identities, activities and discourses in ways that sustain inequalities of power, two political strategies emerge as thinkable. The first seeks to weaken 'proper place' as a normative principle; the second seeks to redefine it. We can see challenges to the significance of proper place in, among other things, constant transgressions of its order, as evidenced by the queer and AIDS activism of the 1990s (Brown 1997; Somella/ Wolfe 1997; Bell and Binnie 2000). Through the generation of surprise, wed-ins, die-ins and kiss-ins sought to trouble norms of 'in placeness' through a theatrical and symbolic politics of critique. Yet while these actions sought to disrupt the exclusion of lesbians and gay men from mainstream life, they did not all seek to challenge notions of the 'proper' per se; nor were actions necessarily concerned with challenging the *legitimacy* of the conduct from which lesbians and gay men had been excluded. While undoing norms of proper place may be advocated by some postmodern scholars, others are less ready to give up all commitment to a normative project on which proper place seems to reside. For proper place does not have to designate reactionary forms of spatial segregation or propriety.

To the extent that it simply accords 'spatial rightness' to particular interactions and conduct, it has a much more open-textured quality.

Thus, alongside troubling proper place, lesbian and gay action has also worked to revise norms of the proper. First, and more narrowly, activism has sought to re-establish where the proper place of lesbians and gay men (and to a lesser extent queers and transgendered folk) might be. In some cases, this has taken a transient form as city landscapes became briefly appropriated through marches and festivals. In other cases, the development of visible, gay, commercial and residential spaces functioned as a more permanent strategy for inclusion within the proper. Second, actions have reframed proper place through articulatory practices that suture it to norms of equality and diversity, decoupling proper place from its usual allies of convention, order and security.

My second normative principle – that of public and private – is best seen as a co-constitutive dyad. Given the vast, sprawling literature on these terms, and on their social and historical development, my comments here will be restricted to those that relate to the public/private as normative principles in relation to same-sex spousal rights. My use of the terms also needs to be seen in the light of the space carved out by the intimate/impersonal (see chapter 3) – a dynamic often identified in the language of the public/private.

In its prevailing application as a normative principle, the term 'public' highlights the necessary ways in which certain spaces, practices, obligations and interactions are organised and (importantly) made legible according to principles of impersonalness, impartiality, openness and disinterest. Underpinning this use of the term 'public' is a tension between two ideas: the common and the strange. Public maps on to and straddles these different ideas, the first with its concern for a unified collective entity, the second with its emphasis on unbidden relations between strangers (see also Calhoun 1999; Cooper 1998b). The tension between them, however, is evident in the difficulties liberalism faces when confronted by the challenge of social diversity. Does public identify the common ground that can continue to unite people despite their differences, or does it refer to the space of irreducible difference that cannot be negotiated away?

These concerns, anchored in a particular reading of the social, have not remained unchallenged. Critics have questioned the extent to which the 'common' ever prevailed; feminist work has been especially influential here in highlighting the extent to which norms of impartiality and shared interest have worked simultaneously to protect and obscure the interests of dominant social forces. Underpinning this approach is the claim that the norms to which the public is ostensibly articulated: impartiality, objectivity, disinterest (see, e.g., Steinberger 1999), obscure the 'anti-norms' – alienation, hierarchy, irresponsibility, exclusion and fear – that really govern public conduct.

Others, meanwhile, have worked from the premise that public norms do exist and are worth protecting, particularly against the (neo-)liberal move towards a minimal public. In the process of arguing for the defence and, indeed, the expansion of public norms, some radical commentators have gone beyond the practice of immanent critique to resuture the public to different norms. From this perspective it is not enough to create a more expansive and embracing public; the norms with which the public is associated also need change and refinement. To the extent that this operates as a counter-normative project, it is not entirely idealist. New normative configurations are not pulled out of thin air, but drawn from the seeds of alternative articulations that exist within the present. Stretching and expanding these tentative connections become the basis for imagining future possibilities which, in turn, offer a critique of how things are 'now'. Multiculturalism and, more recently, cosmopolitanness provide two lenses through which the public has been reinterpreted, articulated to norms of openness, heterogeneity, accountability, stimulation and excitement (see also Bohman 1999).

Intersecting this public is a configuration of elements embraced by the term 'private'. As with public principles, private principles both organise, and are read off from, social life as facts and ideals. In this way, the 'private' circulates through the social body, gaining force from the normative principles it appeals to, while enabling other normative principles, in turn, to draw strength from its presence. In thinking about private as a dominant normative principle, I want to include two different, interconnected meanings: the controlled, differentiated access to knowledge, things, sights, intimacies and places (by oneself or another),[4] and 'akinship' – where belonging, responsibility and identification are underpinned by the convergence of blood, heritage and similarity.

Conventionally, the withholding of knowledge, alongside the imperative that lesbian and gay desire not be witnessed, proved central to the nexus between homosexuality and privacy. An older working of sexual privacy, as a normative principle, anchored the enforced closeting of improper forms of sexual conduct and erotic feeling in norms of appropriateness and self-discipline. Such norms demanded that access to knowing and seeing, in particular, should be restricted. Today, however, this articulation has weakened as sexual privacy has become, to a larger extent, reoriented around the agency of the subject. Although older understandings of privacy in relation to gay sexual expression endure, they confront newer normative practices in which a refusal to know others, and even more oneself, is read as pathological. This version of privacy does not expect sexual information to be unrestricted (see, e.g., Nagel 1998); however, in identifying the limits to knowing, a tension surfaces between allowing and empowering individuals to determine where the boundaries of access to knowledge, decision-making or sight of them should lie (Young 1997b: 162–3), and the rights of 'strangers' not to see, hear or know too much. Norms of privacy embrace both – as revealed by the opposition 'outing' encountered for both taking away the right to disclosure by the individual concerned and for imposing unwanted knowledge on others.

Alongside norms of restricted access, the term private, as I have said, draws attention to principles of 'akinship'. I use akinship to flesh out the space where familiar and familial elide. It is therefore a key term in considering the effects of spousal recognition as I discuss further below. Unlike more conventional, liberal understandings of the 'private sphere', akinship is not anchored in a specific place, such as the home. Rather, it identifies feelings of belonging, identification and comfortability derived from familialising practices, symbols and edifices. Although the familiar can be associated with spaces of control and restriction, to the extent that akinship functions normatively within liberal Western societies, it tends to downplay the unpleasantly familiar. Thus, for many lesbians and gay men in a context of non-recognition or acceptance of their sexuality by parents and relatives (Johnston and Valentine 1995), akinship may be felt more keenly away from 'home', in neighbourhoods where gay, lesbian or queer-identified individuals, venues and interactions are visible.

It should be apparent from my discussion so far that public and private norms do not operate according to a binary division, whereby people, activities and norms are simply and straightforwardly allocated to one side or another. Most spaces and activities combine the two. The complex relationship between public and private is apparent in relation to public sex, a subject which has attracted considerable attention within lesbian and, more particularly gay, studies over recent decades. While sex may take place between people who do not personally know each other and in circumstances where little personal information is divulged (see Murray 1999: 161), strangers may simultaneously be familiar in their physique and repertoire, and in their shared knowledge of the social codes that operate. Sexual spaces may also be coded as private in the sense of being secluded, familiar, exclusionary or amenable to control by those who use them, despite being formally accessible to a wider population and non-privately owned. This synthesis of public and private can be seen in Hollister's (1999: 63–4) discussion of sex in American rest areas which he refers to as a 'collective private sphere'.

Social norms and the pursuit of spousal equality

Proper place and the public/private work to sustain inequalities through exclusions, hierarchies of who or what belongs where, selective access, public benefits, and the grounding of emotional connection and responsibility on the narrow terms of conventional akinship. Yet proper place and the public/private, like other normative organising principles, demonstrate flexibility, unevenness and the capacity to be articulated in more egalitarian ways. I now wish both to develop and concretise this analysis by focusing on lesbian and gay attempts to achieve partnership equality through institutional recognition.

For many proponents, 'gay marriage' is simply the materialisation of current normative understandings of proper place and the public/private. From this perspective, denying recognition to same-sex relationships is anachronistic, an exception no longer warranted. For these marriage advocates, the wider exclusionary or hierarchical implications of reform concerns them little. However, my perspective is different. What I want to consider in the rest of this chapter is the relationship between lesbian and gay marriage reform and the broad project of undoing inequalities, as mediated by the presence and power of the two normative organising principles outlined above.

From the perspective of equality of power, lesbian and gay marriage can be read as a progressive venture (Kaplan 1994). It gives lesbians and gay men access to a structure long denied and, as a result, to some of the economic and social benefits from which heterosexuals as a class have benefited (Chambers 1996). Consequently, it might be argued, reform enables lesbians and gay men better to pursue their own conception of the good life – whether this includes marriage or not (Søland 1998). Yet, framing equality according to a group-based paradigm is, as I argued in chapter 4, also problematic. It suggests that lesbians and gay men have shared interests and needs, and that as a class equality means access to the benefits possessed by groups more privileged than they. These assumptions can be disputed in two primary ways: first, by emphasising the diversity and heterogeneity within lesbian and gay constituencies, and second, by highlighting the importance of normative principles that undercut reading equality as remedying a 'lack'. The discussion that follows is underpinned by these two counter-positions.

Since the early 1990s, the progress of same-sex relationship recognition has moved at such a pace that any attempt to delineate the current state of play becomes immediately out of date. Yet despite the range of approaches taken, we can identify a strikingly high level of global isomorphism. The main techniques for delivering greater recognition of lesbian and gay partnerships have been fivefold. They are judicial finding of a quasi-marital arrangement through widening the meaning of relevant terms such as 'spouse'; de facto recognition of relationships by public and private bodies, such as schools, hospitals, insurance companies, pension plans, employers and private leisure clubs; legislative reform to recognise gay relationships in particular contexts, such as immigration; the introduction of (domestic) partnership status (by city, regional or national government); and state institutionalisation of same-sex marriage (see generally Wintemute and Andenæs 2001; also Goldberg-Hiller 2002).

I want to start with the relationship between same-sex spousal rights (SR) and principles of proper place. In talking about place, however, my use of the term is largely figurative, since I am less concerned with physical spaces than with legal, social and cultural forms. Views about the impact of SR on relationship propriety differ widely. At one end of the spectrum – against the claims of reform advocates that the continuing properness of marriage depends on opening up access to outsiders who rightly belong within – conservative opponents fear that recognising gay relationships will create a new, anti-disciplinary, free-for-all infecting and colonising the privileged terrain of traditional marriage. Left critics, in contrast, fear that SR will create a new, disciplined space which – depending on the terms of lesbian and gay entry – will either orbit marriage as a second-class satellite, or lead to a broadening of the marital terrain. In other words, if the concerns of left-wing critics prove true, same-sex SR will scarcely challenge the character, and even less the authority, of the proper place of marriage; it will simply be colonised by it. Indeed, SR may go further to entrench and fortify not only the proper place of marriage within social life, but in addition the proper place of individuals within it as social and economic benefits, responsibilities and rights are organised around, and work to install, appropriate 'complementary' roles.

I discuss below the extent to which same-sex marriage can work against these tendencies, restructuring relations among couples according to the egalitarian model many lesbians and gay men avow. However, it is important to recognise that not all lesbian and gay advocates of SR desire to promote counter-normative structures, whether in relation to domestic labour, economics or lifestyle. In relation to the latter, conservative gay advocates of marriage and registered partnerships see formal recognition, at least in part, as a way of purifying the space of gay and lesbian affective and sexual practices. For them, SR offer an opportunity to socialise and discipline gay men, while, at the same time, differentiating and separating mature members from the infantile, high-risk and contagious who 'give us all a bad name', and who threaten to sully attempts at creating newly respectable homosexual spaces (Dean 1994). In this way, conservative marriage proponents seek to rework the relationship between community boundary dynamics and norms of proper place – to shift the boundary so that it runs through both gay and heterosexual communities, recognising, in economically and racially coded ways, the mature and immature, the ruly and unruly, responsible and irresponsible in both.

For conservative proponents, the propriety, naturalness and inevitability of the constituencies explicitly excluded, namely the very young and those with spouses already, is clear. At the same time, the exclusion of those who refuse or fail to opt in is legitimised through discourses of choice. But the community boundary dynamics of spousal recognition do not just affect those welcomed or excluded from its ranks; SR also threaten to reclassify and discipline, to break up – through the institutionalisation of rights and duties – continua in lesbian and gay relationships of friends, lovers, 'families of choice' and acquaintances. In this way, what has emerged as a complex unofficial space of blending, subtle movement and evolution within lesbian and gay communities (Weeks et al. 2001: 57) risks being segmented into rigid compartments – with divisions inserted to define proper behaviour and feeling – according to officially established and recognised hierarchies of kinship and commitment.

The logic of marriage and spousal rights provides a technique of governance to redomesticate and retemporalise practices such as homosexuality that became, in the late twentieth century, far more widely imagined and lived out. From this perspective, we can see spousal recognition as an attempt to reharness and contain what undoubtedly was for some an increasingly pluralised and unanchored dynamics of desire, where time no longer determines and allocates proper conduct. This lack of time mapping enabled a freer, more personal crafting of relationships away from the institutional and routinised sequencing of dating, engagement, marriage and children (Weeks et al. 2001: 107). SR thus function as a reassertion of order, where official timekeepers determine and scrutinise the proper phasing of gay relationships. The practical and cultural implications of 'keeping time' for lesbians and gay men, determining when spousal or other recognised status and entitlements (e.g., permission to immigrate) kick in, are impossible to predict. However, spousal equality does appear to exemplify a growing convergence in the way hetero and homo lifestyles are disciplined.

Against proper place?

If SR, particularly as marriage, risk bringing lesbian and gay relationship and kinship structures more closely into line with heterosexual conventions (Brownworth 1996; Card 1996), perhaps, then, the more interesting question is whether SR can also work against incorporation. To what extent can SR challenge the hierarchical distribution and segregation of people, identities and activities? In addressing this question I want to break open SR to suggest that its effects may, to some degree, depend on the choice of strategy: how rights and recognition are argued for, the forms of institutionalisation put into effect and the way in which spousal status is inhabited.

I want to start by considering the arguments made in support of institutional recognition: can SR be advocated in ways that assert its equivalence to heterosexual marriage without assuming the absolute legitimacy of either? This challenge echoes issues raised in relation to 'gays in the military', where too often a pragmatic anti-discriminatory discourse slides into a more patriotic and uncritical valorisation of military practice. If demands for spousal recognition are not to reinscribe gay relationships according to conventional hierarchies of the proper, advocacy needs to affirm other kinds of relationships or personal statuses too, articulating proper place to norms of diversity, consent and equality. For one of the problems with the pursuit of SR, as I have suggested, is the way in which its claims explicitly (or otherwise) trivialise, infantilise or subordinate other relationships. These other relations might include the fleeting sexual encounter with an unknown other – usually pitted as the antithesis of the conjugal couple – as well as friendship networks (Kaplan 1997). But locating SR within a plurality of authorised relationships (see also Warner 1999: 90) also raises questions as to the entitlements to which SR should give rise. While arguments can be made that the proper place for intimate decisions relating to health management or death is sometimes (if not always) with lovers, close friends or household members, claims that conjugal couples should have special rights or access to resources – whether of their partner, the state or the private commercial sector – need to establish, rather than assume, spousal partners' propriety as the place for such material advantages.

The second issue concerns how partnership recognition is put into practice. Can it be given shape in ways that help to dismantle relationship hierarchies: pluralising who and what constitutes the proper place for particular powers, rights and obligations? In other words, does the expansion of spousal recognition open up opportunities for troubling the convergence of responsibility and entitlement in the single, intimate other, to create instead a more heterogenous and diversified response? To explore this further, I want briefly to consider three different forms that institutional recognition by governments and employers might take (see also Eichler 1997): contract, opting in and regimes of default. In practice, these forms often overlap or are combined in particular ways (see Eskridge 2001: 121). However, for ease of discussion I shall deal with them separately. Contract reflects and helps to affirm a particular conception of proper place in several respects. First, the contract itself becomes the 'proper place' for identifying and producing particular entitlements and obligations, binding so long as proper formalities have been followed. Second, the proper place for determining the allocation of commitments becomes located in the parties concerned. The right to draw up a contract emphasises, Weeks and his co-authors argue (2001: 128), 'privately made commitments' rather than ones 'imposed . . . from outside' (although they are still contingent on the 'outside', since effectiveness depends on authoritative recognition). A contractual model therefore has the potential to escape pre-given categories of recognition, for there is no necessary reason why a spouse (same-sex or otherwise) should function as an elective 'next of kin'. It enables individuals instead to decide whom they wish to designate as the proper recipient of various benefits and decision-making powers.

This might be their intimate partner, it might be someone else, or they may choose to spread benefits and powers across different parties. For, again, there is no obvious reason why the person responsible – the proper place – for intimate medical decision-making is the same proper place for pension entitlements on death.

The second approach to recognition, 'opting in' through marriage or institutionalised partnership status, retains a quasi-contractual flavour in the sense that applicants choose, within certain parameters, their spouse (see Card 1996: 12–13; Green 1996; Halvorsen 1998: 216; Søland 1989; Sullivan 1997). Here, a ceremony, utterance or signature may combine to denote the crossing of the threshold into the conjugal unit, although, following heterosexual marriage (O'Donovan 1993), other acts may also become required to consummate the relationship. Kaplan (1994: 353) and some others have advocated opting in over individual contracts on the grounds that it is more financially accessible as an 'off the rack' procedure that does away with the need for expensive, time-consuming formal contracts. At the same time, opting in has several drawbacks as a way of organising institutionalised commitment. Feminist critiques of heterosexual marriage questioned the capacity of less powerful or more dependent parties to consent fully (see Pateman 1988; O'Donovan 1993: 88–9). While gender in same-sex relations does not operate in the same way as a principle of asymmetry, class, age and race, in particular, may take on this role. Moreover, the symbolic power of 'opting in' to a marital or quasi-marital structure accentuates the propriety attached to its rules and assumptions, namely that a wide range of rights and responsibilities should be located with the conjugal partner; that 'improper' selections, such as biological relations, children and multiple partners be excluded; and – to the extent that gay spousal rights are differentiated from heterosexual ones – that certain rights and responsibilities do not follow: for instance lesbian and gay relationships have been deemed in some cases not to be a proper place for children.

The third approach, that of the default regime, takes away explicit choice: we cannot choose for our partner not to count. Instead, governments, courts and, to a lesser degree, employers allocate benefits, powers and obligations according to officially sanctioned conceptions of appropriateness. This may be on the basis of particular relationships or according to other criteria, such as 'best interests'. This third approach has the potential for a more radical, collective revisioning of 'proper place'. For instance, it can avoid the individualist, predictive and voluntarist assumptions particularly apparent in contract, allowing responsibilities to reflect the relationship as it is at a particular point in time rather than being determined by a prior contractual prediction (Young 1997b: 108). Default regimes can also spread responsibilities more widely, such as through extending tort-based duties of care to new parties. It is also, arguably, the most compatible with enhanced state provision of welfare, as state structures allocate and determine powers, responsibilities and resources. Millbank (1998: 130–1) argues that an advantage of presumption-based schemes is that they protect vulnerable parties where the member with more power refuses to 'opt in'. It also, she argues, means that relationships do not have to be publicly performed or announced until they are called on (although some appropriate relationship evidence will be required) (Millbank 1998: 131–2). At the same time, given that governments are usually more conservative than social movements on these issues, the creation of statutory regimes may do little to challenge relationship hierarchies; they may also be based on problematic notions as to who counts and when. As Millbank (1998: 131) argues, a scheme modelled on the middle-class heterosexual lifestyle, for instance where couples are identified through their shared bank account or mortgage, may prove distorting or inappropriate for many lesbians and gay men.

In exploring the capacity of same-sex spousal recognition to contest conventional conceptions of a marital proper place, with the inequalities it sustains and legitimises, the third element to which I wish to draw attention concerns how SR are *inhabited* once in operation. Is there a danger that lesbians and gay men enter through 'marriage' and commitment ceremonies in too sombre and respectful a manner? Would greater levity, parody, pastiche or the explicit incorporation of non-heterosexual elements enable gay marriage to be a space that is not a proper place? Can the spousal domain be one in which the 'out of place' functions for lesbian and gay activists less as the constantly feared intruder – the boundary marker that delineates gay propriety – than as the one whose entry is permitted and even celebrated?

The possibilities for drag weddings, staged non-monogamous commitments, serial registered partnerships, and celebratory divorces clearly invoke a queer transgressive politics; nevertheless, in exploring this as a counter-normative strategy three difficulties immediately emerge. First, why would people enter into an institutionalised arrangement if they disagreed with it? While those entering for purely pragmatic reasons may signal their normative distance from the event through parody, the evidence so far suggests that lesbians and gay men applying for spousal recognition do so in a committed rather than ironic manner. Indeed, parody is often used as a means of challenging the propriety of particular exclusions caused by the dynamics of community boundary maintenance, for instance, the mock, camped-up wedding ceremonies performed in London's Trafalgar Square by same-sex couples to protest against their exclusion from the right to marry. Second, the creation of 'improper' conjugal performances may be too oppositionalist, where the out of place is valorised regardless of what it entails. This tension goes to the heart of a counter-normative politics which seeks to articulate 'proper place' to diversity: namely, are certain activities and identities, such as the eroticisation of violence, adult–child sexual relationships, explicit non-commitment (and forms of emotional 'betrayal') *legitimately* out of place?

The third danger in attempting to disrupt the creation of a proper, legitimate space is that it risks trivialising and ridiculing lesbian and gay relationships while leaving other 'marital' relationships unblemished. Indeed, to the extent that same-sex spousal recognition functions as a discrete satellite form, its parody may strengthen and further naturalise the heterosexual 'original'. It is in response to this that I turn to my final, more general strategy: occupying the space of institutionalised gay relationships in order to challenge and contest the *heterosexual* spousal form. One form this might take – remaining with the politics of parody and disturbance – is a gay 'marriage of *in*convenience'. A second, more outward-looking strategy involves alliances with progressive or radical heterosexuals. While SR might encourage heterosexuals to feel that marriage is modernising and thereby becoming less politically problematic, the development of registered partnerships poses an alternative that heterosexuals might enter too (see, e.g., Eskridge 2001: 120; Young 1997b: 110). While such partnerships are a form of relationship institutionalisation, they lack marriage's historical associations with property, class and gender inequality. Moreover, to the extent that gay and lesbian couples are shunted into this satellite space, heterosexual entry offers a form of solidarity or refusal to partake of a more elite space – a watered down version of heterosexual feminists' earlier, politically driven, repudiation of marriage. Yet what is interesting is the extent to which the statutory creation of some registered partnerships schemes explicitly excludes entry by differently gendered couples (Warner 1999: 126). The continuing illusion of this 'different but equal' approach highlights the ways in which norms of equality and choice collide with hierarchy and convention in the process of resuturing the proper place of marriage.

The organising principle of proper place has proved to be an incredibly significant, although not always explicit, frame for thinking about SR for both proponents and some critics. While conservative opponents argue that the proper place for lesbians and gays is somewhere other than marriage, advocates of recognition argue that they rightfully belong within it. Some queer activists and feminists may challenge the valorisation of the proper, with queer activists, in particular, advocating the improper instead; but this more radical approach is a minority one. For the most part, proper place operates as a primary structuring norm integral to living within, and even to imagining, viable social life.

My discussion so far suggests that proper place organises same-sex spousal recognition in several ways: it shapes the discourses used in argument, the regulatory forms adopted (particularly decisions about who can and cannot participate), and the ways in which SR are inhabited – where a proper crossing of the threshold can range from an ornate, traditional wedding to a carefully crafted, intimate ceremony to a quick, impersonal act of registration. Yet is the relationship between proper place and spousal recognition simply one of absorption and colonisation on the part of the former? Is proper place such a powerful and solidified organising device that it can weld SR to it without being marked or affected in any way in the process?

Developments in the area of lesbian and gay SR suggest two primary forms of impact. First, same-sex spousal recognition has contributed to the articulation of proper place to normative principles of diversity and heterogeneity. While some gay critics argue that same-sex spousal recognition places the survival of other domestic and intimate structures in jeopardy, there is little evidence to support this claim. Rather, opening up SR to enable a broader and more diverse space 'within' allows the authority and insignia of the proper to be bestowed on different sorts of relationships; the consequences of this are considerable. The state's formal acknowledgment of lesbian and gay relationships through SR arguably validates a more heterogenous approach to affective and familial relationships in general that can extend to other policy areas and to main-stream cultural representations. Against this argument it might be claimed that since lesbian and gay marriages (or registered partnerships) are likely to remain numerically insignificant, they are unlikely to revise wider policy or cultural practices. Moreover, to the extent that lesbian and gay relationships impinge on normative principles of proper place in these other contexts, is there any reason or evidence to suggest that marriage or registered partnerships will be the linch-pin of this occurrence? Critics might point to the way in which heterosexual non-marital relationships are increasingly recognised within public policy and mainstream culture. But are lesbian and gay relationships different? Can it be said that their traditionally subordinate or marginalised status undermines their capacity to impact on principles of the proper without *first* being incorporated within the proper?

This is the claim of progressive reformers. Yet, in making it, they confront the critics' rejoinder that same-sex SR, particularly when acquired through new, 'off the rack' arrangements, consolidate and solidify proper place as an organ-ising principle. This raises the second form of impact that SR have. Same-sex SR is a striving for recognition and status – to be deemed proper. Through in-stitutionalisation and formalisation, new areas of social life become translated into classificatory form. The authority of the proper is heightened by procedures which establish whether status has been accorded correctly, particularly where different relationship categories co-exist: for instance, marriage, registered part-nerships, civil unions, domestic partnerships and 'common law' marriage, into which groups of people can be placed. Indeed, one side-effect of the creation of these different arrangements is the possibility of a plethora of legal challenges and case-law consolidating what constitutes a proper entry and dissolution (and between whom).

But if same-sex SR work to consolidate proper place as an organising principle, albeit in a more heterogenous form, what effects, if any, is this likely to have on relations of inequality? I suggested earlier that a major reason for discussing normative principles is the significance of their role in mediating and bridging different inequalities; but can we in any way hazard what impact the convergence of SR and proper place is likely to have? I suggested above that proper place in Western liberal societies no longer works primarily through formalised exclusions and segregations of race or gender. While zoning effects continue to operate in other forms, such as through immigration law and social policy, proper place for the most part largely entails divisions and distinctions based on conduct or role. The extension of spousal recognition to lesbian and gay couples, with its emphasis on choice, preference and the rights and duties this 'voluntary' transition brings, fits snugly within this shift from fixed status to role and conduct.

At the same time, the normalising and obscuring of status-based exclusions generated through spousal recognition echoes similar workings of the proper in other spheres, as I explore in the chapter that follows. These workings of the proper perform – even if the jury is out on whether they secure and reproduce – inequalities of class and disability, in particular, as the indices of 'spouse-like' relations, commitment ceremonies, domesticity, and private responsibility privilege able-bodied middle-class couples (Carrington 1999). But what also gets performed and, we might argue, *reproduced* are inequalities of preference and conduct. In the context of spousal recognition, and depending on the form institutionalisation takes, disadvantaged preferences include serial monogamy, recreational sex, friendships and other complex configurations of intimacy. As I explored in chapter 3, inequalities of preference and conduct are socially inevitable. The question therefore is not whether they should exist but what form they should take. Proper place cannot resolve this question. As a normative principle it works mainly to protect and fortify the norms it becomes sutured to. One cluster of normative principles, closely coupled to proper place, that might provide more of an answer, is the public/private.

From kith to kin

Private norms of akinship dominate Western, liberal societies such as Britain at the turn of the twenty-first century. This privileging has clear implications for equality. Normative principles such as akinship structure relations according to a nexus of kinship/home in which social distance correlates with lowered obligations. In an economically asymmetrical world, this reinforces and legitimates inequalities and descending responsibilities within – but also between – nations and regions. But how does this relate to same-sex spousal recognition? Does gay marriage reinforce akinship's descending spiral of commitment?

Spousal recognition has emerged as a political demand in a social and cultural context in which meaningful life is seen to depend on freely chosen, intimate relations (Kaplan 1997: 209; Weston 1995). According to Jeffrey Weeks and his co-authors (2001: 96), '[I]t is implied that successful cohabitation is an indication of the mature or "real" couple . . . the authenticity of this form of relationship is enshrined in the value afforded living together through domestic partnership policies and legislation.' As I suggested above, spousal rights – in their confirmation of the special status accorded the committed couple – shift the locus of information and decisional autonomy from the individual to the couple, a switch Card (1996) identifies as not unproblematic, since, once two people are unified as spouse, it becomes harder to protect the body or belongings of each from the other. But this is not the only way in which the boundaries are redrawn. I suggested above that spousal status differentiates partners from friends – if not emotionally then at least legally. However, the *social* demotion of friends is also a potential effect of the way in which same-sex marriage (particularly when combined with child-raising) recuperates biological kin. Families of birth who, for many lesbians and gay men, were less important emotionally and practically than close friends (Carrington 1999: ch. 3; Weston 1991), appear to becoming reclaimed as they accept and incorporate gay marriages within their kinship networks (Carrington 1999: 211). This commitment and access to traditional forms of familialism is celebrated by conservative gay activists; Andrew Sullivan (1997), for instance, has defended SR, at least in part, *because* they facilitate acceptance and, hence, belonging within kinship structures.

Quintessentially then, spousal recognition does not mobilise a counter-normative public oriented around strangers (see Cooper 1998b), except inasmuch as the spousal partner has shifted from legal stranger to kin (Mohr 1997: 92). Same-sex marital status, and the nexus it constructs between romantic relationships and legal/economic/social rights and obligations, tips the balance further away from relations with unknown persons. Christine Pierce (1995: 12–13) exemplifies the ambivalence of this, when she suggests that '[u]nfortunately, priority rankings among various kinds of claims are determined by the cultural maps worked out by individual societies, and nearness and kinship are real and important . . . it is important for the sake of creating new sentiments to press for gay marriage so that lesbians and gay men can become visible as . . . families, and kin.' From the introspective space of spousal recognition, the stranger is an outsider to whom less is owed and to whom access is definitionally barred. Indeed, entry in the form of 'marriages of convenience', whether heterosexual or now homosexual, comprise a form of cheating or transgression that, in their cynical advancing of (mutual) self-interest, fundamentally renege on the familial and romantic character of the conjugal space.

The private orientation of spousal recognition has a number of implications for a politics of equality. At an abstract level, it reinforces the idea that little is owed to the stranger *qua* stranger; responsibility is rather to family and kin. This has obvious implications for relations of inequality beyond those of sexuality. Racism, imperialism and ethnic relations are most directly affected by the re-embedding of an akinship which impedes challenges to existing distributions of power and resources by privileging family, proximity and similitude. At the same time, this argument is complicated by claims that anti-familialism universalises the experiences and needs of white gentiles, ignoring the solidarity and connections necessary for black and other minority ethnic people, including lesbians and gay men, that can emerge out of kinship relations (see, e.g., hooks 1990).[5] What is important about this critique of anti-familialism for my discussion here is its premise that public and private norms are interconnected. For while pride, identification, solidarity, empathy and support may come, in a racialised society, from family as much as from lesbian and gay communities, what is important about these kinship relations is the way in which they help to empower black people and others in their relations with strangers. In this sense akinship, to the extent that it organises, and is read off from, relations among less powerful constituencies, can be seen as assisting the pursuit of a more egalitarian public.

While the empowerment that comes from participating within kinship structures is important, embedding responsibility and concern for welfare within the couple itself has other less progressive implications. Susan Boyd (1999) and others, writing in Canada, have suggested that judicial and political support for same-sex spousal recognition there was motivated by the state's desire to privatise social welfare more effectively: a process that depended on recognising, and responding to, the new relationship structures that had emerged (see also Boyd 1996; Boyd and Young 2003).

If traditional forms of kinship, domestic responsibility and highly raked emotional commitment provide the dominant configuration of privacy within which same-sex spousal rights has largely settled itself, what potential is there for same-sex marriage and partnership status to rearticulate private norms? Several authors have suggested that lesbian and gay relationships, by rejecting gender roles (Weeks et al. 2001: 99), enable private norms of selective access to information, decision-making and control to be located within the context of greater household equality and democracy (see also Cox 1997a, b). But this claim has also been criticised and rejected. Carrington (1999: 177, 217) argues, based on his research in the Bay Area around San Francisco, that although lesbian and gay couples often wish to represent their relationship in egalitarian terms, the reality is frequently different. Carrington (1999) adopts a domestic labour approach to private relations in ways that map on to my earlier discussion of the intimate/ impersonal. His argument is that the demands of paid labour, particularly within the commercial sector, structure home life, placing greater domestic responsibility on those whose jobs carry less status and apparent strain, especially within affluent households. Carrington's research demonstrates the difficulties that confront attempts to create more egalitarian domestic lives given current social dynamics. However, his work also reveals some of the ways in which domestic relationships outside the loop of official recognition and regulation can place particular burdens and risks on those with less power, including, in this case, those whose contribution is domestic rather than paid work (Carrington 1999: 207–9).

So far, I have suggested that same-sex spousal recognition may – if it does not strengthen – at least reflect a shift towards private rather than public norms. While this may appear to some degree self-evident, I want to complicate the picture with another perspective: one that sees spousal recognition as fundamentally concerned with the stranger or outsider. In doing so, I bracket the claim that SR offer a means of bringing lesbians and gay men out into public life, a perspective illuminatingly explored by Carl Stychin (2003) in his study of domestic partnership recognition in France.

Regardless of whether intimate relationships gain institutional recognition, they tend to be acknowledged by friends and some family members. While marriage may validate same-sex relationships in the eyes of some kin, at a practical and material level SR largely work to structure the behaviour of impersonal third parties through the obligations placed upon them (Kaplan 1994). These strangers are not subordinate or marginal subjects but those with political and economic power. It is the government, large corporations, the legal system, the mass media and the healthcare system – key institutions in the mobilising of public norms – that are hailed in the formal recognition of lesbian and gay relationships. For it is these entities whose power to bestow or recognise inheritance rights, pension entitlements, insurance benefits, property assets and medical decision-making is at stake. Unlike many earlier revolutionary movements which sought explicitly and purposefully to undermine the power of institutions such as the state – including through a refusal to recognise or hail them – struggles for equality and rights at the turn of the twenty-first century have repositioned the state centre stage, ironically at a moment when its power is being practically undercut by powerful transnational economic actors. A significant aspect of spousal recognition is that it looks to and, in the process, helps to reinforce the discursive authority of the establishment. It is a demand by lesbians and gay men that the establishment see them outside the terms of their traditional interpellation as sick, sinful or inadequate – that, through processes of akinship, familialisation and the readjustment of community boundaries – they be brought into being as respectable citizens of the local, national and global polity.

It is also a demand which goes beyond acknowledging the 'makers and shakers' of public norms to constructing a particular relationship to them. Same-sex spousal recognition asks the state, insurance companies and employers to regulate lesbian and gay relationships ergonomically – to orient policy and law around the contours of relationships as they are lived rather than as they are 'distortedly' presented. Implicit in this demand is the assumption that recognition generates more just forms of regulation. In other words, lesbians and gay men can trust public actors to treat them fairly once they see them – and reflect them back – as they truly are. While different writers have expressed scepticism about this drive to be fully revealed and known, the relationship between disclosure, recognition and what I call 'ergonomic regulation' has received less attention. Yet the state's reproduction and maintenance of existing norms and inequalities raise strategic questions about the identities and lifestyles that constituents, such as lesbians and gay men, may wish to mobilise and present if these are to form the terrain around and across which regulation operates. This is not a call for dishonesty and distortion, but simply to question a primary assumption motivating demands for SR that (valuable) lesbian and gay lives will benefit from transparency in relationships conducted with official and scrutinising forces.

Conclusion

My starting point for this chapter was the premise that struggles to undercut one form of inequality may work to reinforce others, a process I have explored by examining the relationship between law reform and normative principles, focusing on proper place and the public/private. Normative organising principles secure and strengthen dominant social relations largely through the ways in which they protect the status quo. However, as I have sought to stress, such principles are neither rigid nor unified. The flexibility and unevenness of normative organising principles is exemplified by the two under consideration. Both proper place and the public/private organise, and can be read off from, the social in varying and contradictory ways. While dominant normative principles exist, these are subject to change, and face ongoing confrontation from oppositional or counter-normative articulations. Yet some changes may be easier to achieve than others. It may be easier for 'respectable' lesbians and gay men to become embraced within the terms of the proper on the grounds of their 'sameness' than to rearticulate the proper to norms of social diversity and heterogeneity, although attempts to do this can be seen, particularly in urban contexts, in the affirmation and pursuit of the cosmopolitan, with its own exclusions and asymmetries.

A gradual or accommodatory approach may appear the most likely way of engineering change, but it risks reinforcing the power of dominant norms. Pursuit of lesbian and gay spousal recognition provides several instances of this, with its stress on the proper, akinship and privacy – norms strengthened and recharged through the dynamic workings of community boundary formation, structures of desire and the intimate/impersonal (at the same time as norms, such as proper place and the public/private, justify and structure these dynamics in turn). In the case of community boundary drawing, the effects of this interplay are evident in the relocation of some lesbian and gay couples to be part of the socially recognised 'we', while others remain locked out. However, the claim that single adults and people engaged in non-monogamous relationships are even further excluded through same-sex SR needs to be balanced against the counter-claim that bringing some lesbians and gay men more explicitly inside may benefit others too.

Mapping the effects of same-sex SR on prevailing norms – the extent to which the latter become strengthened or ostensibly revised – demands research over the years ahead. From the current vantage point, certain inequalities seem to be exacerbated by the interface of SR and normative principles. In particular, the associations drawn between the proper and respectability, the emphasis on emotional proximity and akinship, and the drive for recognition from powerful institutions suggest a process in which inequalities of class and geopolitical location, in particular, may become both accentuated and further naturalised – discursively constructed as the outcome of a trajectory of personal choices and national ones, respectively.

The effect of SR on gender inequalities, on the other hand, seem more equivocal. On the one hand, lesbian and gay relationships highlight ways in which domestic relationships can be organised away from gender-based roles and responsibilities (e.g., Hunter 1995). At the same time, embedding same-sex relationships more firmly, through institutional recognition and regulation, within the social dynamics of the intimate/impersonal, particularly in relation to upkeep, relationship breakdown and child custody, risks solidifying 'gendered' inequalities within same-sex relationships as couples take on different forms of labour. To the extent that this does not occur, it may be because social inequalities, of class or race, take their place, structuring who fills the work spaces mapped out by the intimate/impersonal. As Carrington (1999: 215) identified in his study, 'many of the affluent lesbigay families create a greater sense of equality between the partners through reliance on the service economy, or in other words, upon the poorly paid labors of others, notably women of color and younger, less-educated gay men and lesbians'.

I have suggested in this chapter, building on my discussion in chapter 3, that inequalities are not just organised according to binaries of social location. Normative principles also create subordinate positions occupied by those who fail or refuse to live in accordance with prevailing norms. Arguably, this is where spousal recognition will have its most intensely felt effects, as non-monogamous relationships, casual sexual partners, celibacy and serial coupledom become officially separated from those couples doing the 'right thing'. This is not just a discursive process, but a deeply material one, affecting – though not always in the same way – immigration entitlements, insurance premiums, personal taxation, state welfare, inheritance, access to certain goods and services, and the right to make, and be subject to, other's decisions. But, as I have argued, normative inequalities are inevitable. The question is on what basis and according to which norms should they operate? Proper place gives us little help in this regard; it functions largely as a shell or weight, supporting other norms rather than filling in their content. I want to take the point made by Laclau and Mouffe (2001: 188) in relation to democracy but equally applicable here that proper place is 'incapable of founding a nodal point . . . around which the social fabric can be reconstituted'. Public and private may be more useful. Although by themselves they also tell us little, articulated to other normative principles public and private contribute to a richer sense of the 'good' society – whether it is one organised around personal autonomy, an akinship grounded in the ancestral family tree, or an outward orientation that focuses on less powerful others.

In the chapter that follows, I continue my discussion of the role norms play in securing social inequality, focusing on the prism offered by nuisance. Exploring nuisance as a discursive utterance, policy target and mode of regulation, I examine the norms and relationships that it refracts and secures. At the same time, I am interested in the possibility of flipping nuisance over to provide a way of contesting conservative social norms. The chapter addresses this issue in two ways, by exploring the politics of causing a nuisance through engaging in disruptive and transgressive conduct, and by considering the capacity of nuisance to act as a spoke for utterances that consolidate rather than undermine progressive norms and relations.

Notes

1. Principles of inequality, such as gender, also incorporate normative elements; however, they are not *simply* or only normative. As I discuss in chapter 3, the normative or disciplinary effects of gender arise from the particular ways in which modes of power define and saturate it (see also Cooper 1995a). In contrast, the principles explored here are first and foremost normative. As such, they establish the terms upon which

the normative aspects of gender or age, for example, operate. My thanks to Margaret Davies for raising this point.

2. Elsewhere, I have described such principles as normative-epistemological to highlight both elements as well as the connections between them. However, for ease of reading I am referring to them here as normative principles on the basis that the term 'normative' can embrace the ways in which the 'good' *qua* 'right' society is underpinned by ways of knowing and the construction of the 'true'.

3. Normative principles also play a crucial epistemological role in making other forms of knowledge possible. In the context of liberal society, they not only play a central role in allowing society to be known, but principles such as liberty, discipline and consent are also perceived as formative to valid practices of knowing more generally.

4. Whether access is controlled by oneself or others frequently depends on socioeconomic class. So, for instance, the private spaces of the wealthy are defined as desirable and as legitimately secluded from the view and bodies of outsiders. In contrast, the spaces of the poor may be privatised by external forces and processes – particularly in the sense of being shielded from view and entry – to keep undesirable bodies, sights and smells from penetrating and spoiling more prosperous lives and spaces (see also chapter 6).

5. Similar arguments have been made about the role of akinship in working-class communities. According to Carrington (1999: 119), the less affluent households he researched tended to adopt more traditional and biologically driven conceptions of family.

6

Getting in the way: the social power of nuisance

In this chapter I continue my exploration of the relationship between social norms and organising principles of inequality. Chapter 5 considered this interface through the lens of law reform. It examined the ways in which the legal pursuit of greater gay equality was structured in its meaning and effects by norms of proper place and the public/private divide. Here, I want to continue with this critique. Moving on from law reform, I focus in this chapter on the role played by particular injury-based discourses and remedies.

Discourses of injury are central to the wider themes of this book, namely the relationship between equality and diversity politics. What counts as harm structures the forms of diversity seen as legitimate subjects for equality. In other words, for many, 'causing harm' is the basis for distinguishing smokers' struggles against discrimination from those of women. Yet, this confident distinction is not as easy to sustain as it first appears. What we think of as harmful in the first place is structured by social inequalities and values; it is not a neutral category. This can be seen in the repertoire of 'multicultural debates' played out over circumcision, veiling and religious slaughter. It can also be seen in the conflicts between gay activists and orthodox Christians over employment rights. Do Christian organisations opposed to homosexuality 'harm' gay men and lesbians when they refuse to employ them? Are Christian bodies 'harmed' by being legally forced to employ people who live in ways that they fundamentally abhor?

Within diversity politics, harm functions as an essential, benevolent structuring device – differentiating good from evil, signalling the worthwhile and valuable, and establishing parameters for choice and freedom. However, I want to take a critical approach to harm by focusing on the way it both sustains and is underpinned by key norms and values. This is evident in the different, sometimes overlapping, approaches to harm that have emerged within the space of diversity politics. The first associates harm with group domination and oppression. Harm becomes a shorthand for those forms of conduct or decisions which

118

undermine access to resources, capabilities or power by a subordinate social constituency. From this perspective diversity, in the sense of the varied conduct of the powerful, becomes harmful when it impedes equality. The second sees harm in impediments to freedom. Influenced by liberal thinking (Mill 1929; see also Simester and von Hirsch 2002), this strain of diversity politics treats as harmful those provisions, rules and policies which, without good cause, limit or restrict people's varied pursuit of the 'good life'. Thus from this perspective equality can become harmful when it impedes freedom. The third cluster of responses takes a completely different tack by arguing that it is discourses and claims of injury that are themselves harmful. As I discussed in chapter 2, this approach has been prominently applied to claims of cultural or symbolic injury caused by sexually explicit material and fetish-clothing, although Judith Butler (1997a) also makes a case for a more mediated understanding of the relationship between harm and hate-speech. From this perspective, safeguarding both diversity and equality can prove injurious when over-framed as protection from harm.

Despite their differences, one quality these positions share is their tendency to naturalise the values underpinning their respective conceptions of harm. My aim in this chapter is to do the reverse. By focusing on the power of injury-based claims, I want to explore the norms, values and relations they normalise and help to consolidate. In other words, I want to open up and interrogate the links between harm claims, social norms and inequality rather than foreclosing them. At the same time, my intention is not only critical. I do not want simply to interrogate and deconstruct the norms and relations underlying injury-based utterances. While I want to highlight the power of such utterances – particularly when they receive legal recognition – to undermine equality, the discourses underlying such utterances also offer tools that can be deployed against the status quo. But how far does such a plasticity extend? What contribution do injury-based discourses make to an equality politics?

To explore the politics of harm in more detail, the chapter focuses on one particular injury-based structure. Injury, offence, annoyance, hurt are not transparent concepts, nor are they interchangable. They refract social reality in complex and differing ways. Here, the concept I centre on is nuisance – a term which covers considerable ground from mere annoyance to harassment and physical incapacity. Nuisance is an important but under-theorised term for thinking about diversity and equality. First, it tends to be used against marginalised and less powerful social constituencies rather than against those with social and economic status. Second, nuisance has strongly normalising tendencies, which effectively deface those against whom its claims are made. Third, it provides a politically significant legal device that penalises particular acts, presences and identities. Finally, to the extent that the term nuisance provides a way of framing practices that impede and disrupt dominant social processes, adopting the mantle of nuisance has offered one particular basis for political conduct. As I discuss further below, nuisance provides an interesting instance of political and social activism, given its relationship to the embodied, the sensory and the practical.

In order to explore the ways in which nuisance operates, I consider its application in a range of contexts. In particular, I approach the question of how nuisance sustains the norms and social inequalities that underpin it in three ways: by focusing on nuisance as a legal structure, as a public policy dilemma and as a discursive utterance. In their different ways, these uses of nuisance install, impose, authorise, validate and normalise certain ways of living and being over others.

As a legal term of private tort law, nuisance denotes the substantial and unacceptable transmission of odours, sounds, sights and vibrations across property boundaries where the injury emerges from, or is caused by, land not controlled by the victim. '[N]uisance can involve actual physical damage to property . . . and/or non-physical "amenity" damage such as noise or smells which unreasonably affect the plaintiff's comfort and enjoyment of her property' (Conaghan and Mansell 1999: 126) or which interfere with her rights (see, e.g., Newark 1949).[1] As a criminal law term, public nuisance refers to tortious transmissions that detrimentally affect a significant section of a community, to blockages of the highway and to other practices deemed to harm the health, safety or public morals of a substantial number of people within a community (see Spencer 1989; Trachtman 1983). 'The essential character of public nuisance is the infringement of a public right, an interference with the lives and activities of a community rather than an individual' (Conaghan and Mansell 1999: 128). While public nuisance has provided a catch-all class for a range of activities, much of it has now been converted into statutory offences, particularly in the fields of environmental pollution (Steele 1995), refuse (Crane 2000), noise and public order (Parpworth 1995).

Within the field of social policy, nuisance, or antisocial behaviour, has been used to designate a spectrum of activity from spitting, swearing, graffiti and loud noise to harassment, name-calling and petty violence that became coupled, within British policy discourse, with the conduct of poor people in shared urban areas and with tenancies in inner-city neighbourhoods in the late 1990s and at the turn of the twenty-first century.[2] While this largely operated at a governmental level, it both intensified and was anchored in nuisance's application as a term of everyday discourse to describe familiar repetitive impediments or annoyances that distracted people from their activities, slowed them down or simply got in their way.

In exploring how injury-based concepts, such as nuisance, refract the social norms that underpin them, my discussion builds on the previous chapter by focusing on two normative principles: freedom and proper place. As there, my concern is not simply with such principles for their own sake, but also to investigate their role in maintaining social inequality. The ways in which particular configurations of freedom and proper place underpin nuisance, and the impact this has on existing inequalities, form the heart of my discussion in the first two sections of this chapter. At the same time, given the overarching interest in this book in politically driven social change, my interest is also in the capacity of nuisance to be deployed *against* dominant social relations. I explore this capacity in several ways. In the first two sections I consider some of the ways in which nuisance is used to protect the position and entitlements of subordinate constituencies, and the way in which it adapts and changes as established practices come under challenge. Subsequently I explore the extent to which nuisance offers a location that forces seeking to disrupt current norms and relations can productively inhabit. I end the chapter by returning to more conceptual terrain, namely whether nuisance as a discursive framework can be re-envisaged to support other social norms. In addressing this question, I explore the relationship nuisance has to both a counter-normative and anti-normative politics.

Social norms underpinning nuisance

In chapter 4 I suggested that we should approach equality from two angles: the relations of inequality that render social capacity uneven, and the aspiration, albeit unattainable, for individual equality of power. This unattainability is not due solely to present-day conditions that position people asymmetrically. Equality is more fundamentally elusive for reasons discussed in chapter 3. There, I argued that people's pursuit of differing visions of the good life would remain uneven, since societies invariably – inevitably – support and enable certain activities and choices, while others are discouraged, culturally voided or rendered costly, even when they are not explicitly condemned. Nuisance is one device for structuring this difference – as a discourse of media, policy and everyday usage and as a legal framework. But what structuring effects do claims of nuisance have? What social relations and choices do such claims prefer?

I want to start this discussion with a normative principle that has proved central to nuisance discourse and law. It concerns negative freedom, that is freedom *from* the external obstacles or impediments that get placed in the way of pursuing one's interests (see also Hirschmann 1996: 53). According to Isaiah Berlin (1991: 34), negative freedom represents that area in which someone can act unobstructed by others. The defence of liberty thus consists in the negative goal of warding off interference (Berlin 1991: 40). Later in this chapter I consider whether nuisance can extend beyond negative freedom. For, despite its refraction of property owners' positive duties (see, e.g., Markesinis 1989), in its more common legal application and everyday discursive usage, nuisance identifies practices which restrict, obstruct or interfere with an individual or organisation's pursuit of desired acts or goals. So, 'begging with the intent of intimidating another person into giving money or goods, or intentionally block- ing passage by another person so as to require that person to take evasive action to avoid physical contact' (Kelling and Coles 1996: 214–15) are commonly depicted as a nuisance. In contrast, a refusal to act, for instance to give money to someone busking, is rarely defined as a nuisance as is a lack of shelter for homeless people. While positive forms of freedom describe activities that *assist* others to pursue their goals, their absence is rarely identified (except perhaps by those affected) as a nuisance.

To say that nuisance is articulated to negative freedom is, however, not enough. As Waldron (1991) explores in his discussion of homelessness and freedom, negative freedom is not simply a freedom from all obstacles. Property laws, for instance, substantially impede the capacity of the propertyless to pur- sue their interests, but are rarely defined as an impediment to their freedom, and hence a nuisance. To flesh out the normative principles that nuisance refracts, we need to consider the citational and referential chains – that is those principles explicitly hailed as well as those that ground intelligibility – that link negative freedom to other norms.

Nuisance discourse refracts a negative freedom anchored in (and, in turn, helping to support) normative principles of entitlement, legitimate interest and purposive conduct (see also Feinberg 1985: 38). The significance of entitlement is evident if we take the examples of begging and 'cottaging' (sex in public lavatories) being interrupted – by police officers in the first case and others wishing to use the urinals in the second. In both cases, what counts as a nuisance is open to contrasting interpretations: that it is the beggar or police officer who is the obstacle in the first, the sexual participant or prospective urinal user in the second. But, in each case, public discourse is clear: one use is seen as entitled, the other is not. The relationship between entitlement and nuisance is not always as clear-cut. For instance, increasing opposition to people wearing perfume at work in North America has led to attempts to redefine it as a nuisance on the grounds that it interferes with people's entitlement to reasonable employment conditions. This stance, however, has encountered equally ardent objections from those who claim that the attempt to increase regulation and reduce personal freedom, in relation to conduct that harms and restricts no one, is itself a nuisance. While the status of perfume worn at work has become an ambiguous one in certain jurisdictions, the conflict demonstrates both the ability of nuisance discourse to attach to new forms of illegitimate conduct and its capacity to be deployed in the process of redefining what entitlement consists of.

The second principle to have played a significant role in structuring what gets defined as an interference within nuisance law is that of legitimate self-interest. Emanations such as loud music or letting off guns have been declared in a series of cases to constitute a legal nuisance when motivated by spite or malice even where the action was in response to a prior 'obstruction'. In other cases, by contrast, nuisance was held not to exist where the courts ruled that the obstruction was in 'lawful' pursuit of the party's economic or professional self-interest (see, e.g., Cross 1995; Fridman 1954; Wightman 1998: 872–3). As Fridman (1954: 588) suggests, in discussing the court's approach in *Pickles*, a case concerning the withdrawal of percolating water from under a neighbour's land, 'a man who used part of his own land . . . for the purpose of advancing or protecting his economic interests, was acting not unlawfully but justifiably'.[3]

In cases of public nuisance or antisocial conduct, freedom for pedestrians and cars to move instrumentally, purposively and quietly through shared outdoor spaces has been privileged over the freedom to lie, sit or linger and draw strangers into conversation (except perhaps in cases of emergency). Loitering in open spaces, especially in Walzer's (1986) 'single-minded spaces', is not seen as something to safeguard from interference but rather is regulated (and impeded) in a way that contrasts with the response to loitering in one's own private domain (which the law for the most part protects). Yet loitering is an interesting species of nuisance conduct (see Trosch 1993). On the one hand, it identifies and is condemned for its suggested lack of purpose – groups of bored kids, for example, hanging about – which does not bestow any entitlement to freedom from interference. At the same time, loitering tends to be linked, in the legal and popular imagination, to unlawful or stigmatised activities or to conduct-based statuses: prostitution, harassment, begging and gang membership (see Livingston 1997: 555–6). In this way, while the mere presence of largely stationary people may be deemed to lack value, and to impede the capacity of other people to use open spaces instrumentally, the wider fear or anxiety is what people hanging about with no ostensible authorised purpose (such as watching legal street theatre or sitting at an outdoor restaurant) may actually do. In this sense, nuisance claims are about the right to freedom from anxiety, an invoking of the claimant's entitlement to mental space, as I go on to discuss.

Turning mental space

While nuisance claims embrace a range of different kinds of freedom-limiting actions, I want to focus on one distinctive aspect of the relationship between freedom and nuisance. In British case-law doctrine, central to private law nuisance, particularly where the harm concerns amenity rather than physical damage, is the suffering plaintiff. In other words, the question is not simply whether formal rights have been breached, but whether the rights-holder has been detrimentally affected. In popular parlance, nuisance requires not so much a particular kind of response as a 'turning'. This is not the mental turning associated with the pleasurably strange and unexpected, but a turning towards that which, although familiar, is nevertheless undesired.

Identifying demands for attention as a nuisance have most currency when such demands interfere with one of three mental processes: the domestic mindset and equilibrium associated with 'normal' household activities (especially sleeping, cleaning and enjoying one's garden); the focused attention and concentration associated with particular occupations or forms of employment; and the cognitive state deemed to govern and direct legitimate instrumental action in public spaces (such as getting to work, driving or shopping). When people complain about the nuisance caused by the car ahead driven too slowly, the weeds entering from a neighbour's garden, the picket outside their place of work or their neighbour's incessant playing of a single music track, their claim is anchored in their objection to the fact that, for a period of time, their attention has been unwillingly diverted.

How someone is hailed may prove relevant; nuisance claims, especially legal ones, can be sensitive to the differences between a defendant's polite request, and a shout, abuse or intimidation.[4] Also relevant may be the character of the diverted attention. Ellickson (1973: 734) refers to US cases where funeral parlours in residential areas were held to be a nuisance on the grounds that they inspired feelings of depression in local residents through reminding them of death. However, in general, nuisance claims do not depend on the actual content of the victim's lost thought or the character of their response. Excepting cases of retaliation or revenge, hailing distractions as a nuisance is not contingent on whether the 'victim' seeks amelioration, avoidance, an oral venting of their frustrations or appeasement. The main thing is that their attention has been *unwillingly* diverted.

At the core of nuisance's mental quality, then, is the unbidden nature of the distraction. Yet not all unchosen demands for attention have discursive currency as nuisance. Not all people in all situations – youths 'loitering', homeless adults sleeping on park benches or in office doorways, for instance – possess an entitlement to be left undisturbed from the claims and demands of others. Meanwhile some people, such as police officers, have the right to hail all of us, in varying terms, and on occasions to coerce our attention. Whether demands for attention constitute a nuisance in the sense of interfering with our negative freedom depend on whether or not they are seen as socially legitimate. Nuisance utterances, and the norms that underpin them, are anchored in relationships between people that are socially embedded and socially driven.[5]

How context and relationship structure what gets defined as an interference with mental freedom is evident in the following university-based example. A student's conduct may be read as pestering or a nuisance to a professor when demands for attention are made too frequently or outside accepted hours, or adopt an unacceptable form. What constitutes these will depend on the way in which wider conventions become inflected by local (not necessarily explicit) agreement, but might include telephoning the professor at home at the weekend, repeated e-mail enquiries, or entering a lecturer's office to ask questions without an appointment. Whether or not the student is aware that they have transgressed may be seen as irrelevant to the nuisance they cause; however, the mental status of the one who causes a nuisance does come indirectly to the fore, since nuisance depends on repetition. This could involve the same student's repeated failure to pick up cues that his demands are unwelcome, or a situation where a number of students replicate the same act, suggesting either that conventions of appropriate conduct are not shared by staff and students or that they are being recklessly or intentionally broken.

The depiction of students as a nuisance can be read as a depressing consequence of the growing pressures on academics, in Britain and elsewhere, as a result of a steep climb in student numbers unmatched by extra resources. However, the scenario which better illuminates the relational character of how obstacles to mental freedom are understood is the reverse situation, where the professor telephones the student out of hours, e-mails excessively or appears outside her home. Because of the nature of the faculty–student relationship, student dismay at being telephoned – to enquire about missed classes or late essays – may not be seen by others as an infringement of her mental freedom. But what if the professor's actions went beyond this? When does she pose a nuisance to her student in a way that would be generally regarded as such? Later in this chapter I briefly consider the relationship between power and nuisance within the workplace. What I want to highlight here, however, is not only the different, socially contingent thresholds that come into play before unwelcome attention becomes generally perceived as a nuisance, but the ways in which such thresholds occupy a place along a spectrum which ranges from legitimate demands on the one hand to harassment, mental abuse and intimidation on the other. For actors with little social or institutional power, much of their conduct is likely to be read as a nuisance: an unjustifiable, but relatively unthreatening infringement of mental freedom. However, for more powerful actors, their conduct is more likely to fall outside nuisance's band. As I discuss below, the demands of the powerful tend not to be read as illegitimately (even if temporarily) constraining the mental space of less powerful others. For such demands are usually seen as legitimate, given the contexts in which they arise. At the same time, when the demands of people with greater status and power are read as unacceptable, their capacity to inflict harm – that is to injure rather than obstruct the mental freedom of others – tends to take them outside the band of nuisance. In part, for this reason, what constitutes unacceptable demands or conduct by powerful actors is subject to far more dispute and controversy.

The association of nuisance with the unwelcome attentions or demands of others may seem trivial in its signalling of a temporary obstruction; however, its discursive power is far more significant. In particular, it presents a view of the subject as one who does not welcome the demands of strangers (since these demands are by their very nature largely unchosen). While negative attention may be paid to those who hail us without our permission, this is not the attention underpinned by responsibility or concern for the other that social theorists such as Bauman (1993) advocate.

Classifying people as a nuisance expresses a relationship which is antagonistic, where people, and their activities, impede the pursuit of our mental concerns and priorities. In other words, and fundamentally, they do not comprise them. The perception that outsiders are nuisances reduces them to obstacles for which we owe little or no responsibility – a strategy that has been particularly evident in Western countries since the late 1990s. In cases of homelessness, loitering, truanting, drug-taking, or sex work, namely in those contexts where social inequalities are at their most profound or urgent, the prevailing social response has become one of redefining behaviour and people as a nuisance, as impeding, hence extraneous to our goals (see Ellickson 2001).

But it is not only the claims of strangers that get read as nuisance. Kin can also be described as such, epitomised in English popular culture in the portrayal of in-laws. However (and this familial usage also reveals it), the power of nuisance claims comes from its construction of others as outside the scope of our interests, even if only temporarily. For this reason, nuisance utterances within a family are usually constrained. Parents may declare their children a nuisance in the sense of demanding excessive attention or causing annoyance at a particular moment, but this does not usually carry the deeper meaning that their children are merely obstacles, impeding their parents' mental freedom. Indeed, in cases where this is what parents mean, such parents are likely to be perceived as poor, irresponsible carers. In other words, normative parenting requires children to be seen as central to the inner life and construction of interests that parents pursue.

Proper place

So far I have explored the ways in which nuisance, as a discursive and regulatory framework, helps to sustain a negative conception of freedom in which propertied mental space plays a central part. This notion of freedom, however, does not exist alone; it is rendered intelligible by its articulation to, and citing of, other normative principles, especially purposive conduct, entitlement and legitimate self-interest, all of which work to reinforce particular relationships between unequally positioned social subjects. One normative principle with which negative freedom is particularly closely coupled in the nuisance context is that of proper place, a concept discussed in some detail in chapter 5. There, I explored the ways in which proper place, in its dominant form, structured the pursuit and attainment of same-sex spousal recognition. At the same time, I argued, lesbian and gay marriage – in however small a way – had the potential to trouble and revision what proper place entailed.

Proper place is also a central organising principle for nuisance. I suggested above that what counts as nuisance is influenced by notions of purposive conduct; but whether such conduct is authorised as a ground from which to make nuisance claims, or whether it is in fact seen itself as a nuisance, depends on its place-based status. Norms of proper place work, in historically contingent ways, to allocate forms of purposive conduct between the categories of causing or being vulnerable to the nuisances of others. The amenity nuisances caused by activities that produce noise, vibrations, dust and odours depend on being defined as in the wrong place. Successful nuisance cases have produced injunctions and compensation where sporting and musical pursuits cause unreasonable noise to adjacent residents, and where neighbours are affected by unacceptable vibrations or smells from industrial activities. Tort law decisions that declare behaviour appropriate in one area as unacceptable in another demonstrate the zoning effects of private nuisance law (see also Brenner 1974: 406, 414). However, private law largely assumes the abstract or placeless legitimacy of the scrutinised conduct (though see Fridman 1954 on the 'spite' cases); the question is rather whether, given a conflict of adjacent land uses, it is legitimate in its particular location. In contrast, in many cases of public nuisances – rowdy tenants, gang activity, or sex establishments, for instance – concerns about their more ubiquitous inappropriateness may mean that the law is used to zone them out of existence altogether (see, e.g., Boga 1994: 492; Bone 1986: 1141).[6]

The out-zoning achieved by nuisance law, anchored in organising principles of proper place, not only re-establishes the viability of normative spatial allocations but also impacts in varying ways on particular statuses and identities. In some cases, public nuisances may involve conduct only contingently coupled to a particular class of persons. In other cases, the association is a tighter one, where conduct stands in for status. For instance, the eviction of 'nuisance' neighbours or the criminalisation of loitering may target conduct as a more legitimate way of regulating and penalising particular groups of people (see Kelling and Coles 1996: 51). In Britain and the United States, this form of regulation intensified throughout the 1990s as more activities in shared outdoor spaces – particularly those of 'beggars, rough sleepers, peddlars, buskers and others who are threatening or who engage in antisocial behaviour in public places' (Office of the Deputy Prime Minister 2002: 56) – became regulated through nuisance laws. Such measures were accompanied by a discourse of reclamation in which the conduct and people to be displaced were portrayed as 'increasingly brazen', having invaded, infected or taken control of 'public' spaces (Livingston 1997). But there are also some situations in which conduct and identity are so closely coupled or collapsed into each other that it becomes impossible to recuperate a people from the out-zoning of their conduct or 'lifestyle' (see also Campbell 1995; MacLaughlin 1998; Mitchell 1997; Okely 1983). For instance, as Waldron (1991) discusses in his analysis of freedom and poverty laws, if people without private spaces cannot use public ones to engage in the essential human activities of eating, urinating and sleeping, where can they exist? The erasure of classes of people, collapsed into the impropriety of their conduct, is evident in British judicial discussions of the supposed nuisance caused by gypsies and travellers to landowners.[7] Legal dicta describe travelling peoples as an 'emanation',[8] 'offensive to those who have fixed homes',[9] and equate gypsies with 'pestilential rubbish'[10] and unnatural and dangerous uses of the land.[11]

In the process of defining emissions, conduct or status as a nuisance, a particular conception of space is brought into being. This may be a space folded into new configurations of proximity (and distance) as, for instance, when policies of foreign governments are defined domestically as a nuisance. More commonly, the out of place quality that nuisance claims evoke works to constitute, define and explain distinct landscapes, as when the out of placeness of rubbish and litter on city streets functions as the origin and cause of 'vermin and disease' (Office of the Deputy Prime Minister 2002: 12). This process of place-making through nuisance law and discourse is particularly evident in the case of transitional landscapes moving towards gentrification (see also Cresswell 1996: 4–5; Mitchell 2001).[12] MacLaughlin (1998: 428) makes a similar point in his discussion of modernisation and its impact on the place of Irish travelling communities: 'In the branch-plant economy and corporate society of modern Ireland Travellers have also been demonized and constructed as social anachronisms. Thus they went from being a distinct social *and functioning unit* in traditional Irish society to being a social nuisance in the political and economic landscape of modern Ireland.'

Yet, in considering what is taken to be out of place, I do not want to overstate the degree of closure that exists. Not only do counter-normative principles of proper place exist, embedded, for instance, in alternative communes and communities (see chapter 8), but the everyday construction of proper place is frequently subject to contestation and revisioning (see also chapter 5). Oppositional discourses, rhetorically and strategically, construct counter-claims of proper place to denigrate powerful identities. Recent attempts to designate the British monarchy as being out of place offer one example. This manipulation of norms of the proper works on several levels. It includes representations of the monarch as out of place given conditions of modernity and in the context of evolving antithetical national identities, such as that of Australia. It also includes the tactical redefining of royalty's visits to places of work, health treatment and leisure facilities as a nuisance. Dominant understandings treat royalty's presence in our everyday spaces as a privilege, treat or surprise. To the extent that people remain willing to redesign their ends around, rather than in the face of, royalty's presence, their visits are not a nuisance. Rather, it is the action of opponents, their breach of protocol, that is defined as improper. In contrast, to see royal visits as a nuisance requires the affirmation of other ends – that the spaces inhabited are for working, receiving medical attention, or playing sports. From this perspective, it is the demands and presence of royalty that become reframed as out of place.

Consolidating inequality

As should be apparent from my discussion so far, normative organising principles work through law, policy and discursive claims of nuisance to consolidate and reinforce unequal social statuses. But this is not the only way in which the normative principles refracted by nuisance embed relations of inequality. In the discussion that follows, I want to highlight three other means; they concern the interplay of social dynamics, binarised cultural values and institutional relations.

In chapter 3 I argued that social dynamics such as the intimate/impersonal, community boundary mechanisms, desire and capitalism are key means of reinforcing relations of inequality. What gets counted, for instance, as a legal nuisance, through judicial interpretation of freedom, reasonable use, and responsibility, and the way this has historically evolved, is firmly anchored in capitalist processes and relations (see also Brenner 1973; Conaghan and Mansell 1999). But capitalism is not the only social dynamic that nuisance refracts and consolidates. A second is the intimate/impersonal (see chapter 3). Nuisance is frequently associated with the expression of intimate norms in spaces governed by impersonal ones. According to Waldron (1991: 301), 'Since the public and private are complementary, the activities performed in public are to be the complement of those appropriately performed in private.' The mobilisation of norms of inappropriateness, particularly evident in claims of offence, cloak the seepage into public of practices associated with non-dominant constituencies, such as same-sex intimacies, washing, cleaning or urination.

Antisocial behaviour is also typically about doing injurious things in commonly used or public spaces which may be perfectly acceptable when carried out at home. Boga (1994) makes a similar point when he contrasts the pejorative attitudes and legal remedies used in the United States against street gangs with the far more sympathetic response encountered by college fraternities. According to Boga (1994: 488) the main difference is that the latter carry out their actions in designated housing or other 'private' spaces. In Britain, to the extent that antisocial behaviour (or nuisance) refers to acts carried out *within* domestic or private spaces, the harm usually relates to people elsewhere (see also Skogan 1990: 46). Thus, under the Crime and Disorder Act 1998, s. 1, the harassment, alarm or distress necessary for an Anti-Social Behaviour Order (ASBO) to be issued must to be towards persons *not* of the same household.[13] The exclusion of household members from such regulatory structures demonstrates the ways in which these impersonal formal techniques of conflict resolution and remedy are not deemed necessary or suitable for use between intimate family members. While traditionally those holding domestic power were not seen as a nuisance – that is as making illegitimate demands on those below them (husbands to wives, parents to children), the annoyances and infractions of subordinate household members (servants, children, wives) both could and were expected to be dealt with within the ambit of household authority.

The second way in which norms of proper place and negative freedom (as well as of responsibility) are mobilised to consolidate organising principles of inequality concerns the cultural imagery nuisance law and utterances invoke. In chapter 3 I argued that an important aspect of principles of inequality was the way in which they permeated and shaped other aspects of the social. My argument there was that one key reason why it would be premature (at best) to define smoking as an organising principle was due to the weak impact it had on wider norms, epistemologies and social and institutional processes. In the case of nuisance discourse and law, its role in converting inequalities of race and class into sensory experiences of distaste and disgust has been historically considerable (see, e.g., Corbin 1986). Nuisance also offers a means of channelling principles of inequality's more metaphorical associations. I want briefly to illustrate this process, taking gender as an example. A central organising metaphor through which nuisance operates is the relationship between permeability on the one hand and penetration and boundedness on the other. As several writers have explored, these terms are highly gendered (see, e.g., Shildrick 1997:15). While (heterosexual) masculine bodies are constituted as impermeable and penetrating, feminised bodies, particularly when white and middle-class, are depicted as highly porous, unbounded and vulnerable (see also Thomson 2001). '[W]hile the standard (male) body is defined by its intactness, its wholeness, its completeness, the non-standard (female) body is defined by its gaps, its openings, its incompleteness' (Naffine 1997: 88). This division survives even in cases where men's bodies are ruptured, provided that this occurs in legitimate ways. 'Destruction of the male body will not necessarily be perceived to damage the wholeness of the legal body . . . so long as it is thought not to feminise the man, not to reduce him to the ambiguous and fluid bodily status of woman' (Naffine 1997: 87; see also Stychin 1995).

To the extent that gendered associations of the body articulate highly normalised, cultural assumptions, particularly in relation to white, middle-class bodies, they are able to influence how we view other contexts and practices (see also Stychin 1999: 157). At the same time, the wider mobilisation of this imagery works to solidify gendered assumptions. Nuisance, in its private legal form, reads freedom, responsibility and proper place through a narrative of permeability as dust, odours, vibrations, noises and sights travel across property boundaries. Indeed, private law nuisance is grounded in the feminised incapacity of propertied land to erect fully nonporous boundaries or fences to protect itself from sully or contagion, despite the assumptions of liberalism (Nedelsky 1991), and despite a history of ongoing attempts to do so. At the same time, legal nuisance claims present the property from which the nuisance emanates as unable to control and contain its fluids and productive processes, literally expressed in landowners' liability for travellers who urinate and defecate on nearby property. Nuisance law thus echoes historical and, to a lesser extent, current cultural associations of femininity: 'Women, unlike the self-contained and self-containing men, leaked . . . [their] menstrual blood was seen as the internal build-up of noxious waste material' (Shildrick 1997: 34).

Yet, on another reading, causing nuisance is masculinised. While antisocial behaviour is fairly explicitly associated with the obstructions, interruptions and irresponsibility of working-class masculinity, some forms of private law nuisance have more positive connotations – the 'alpha-male' body unable to contain its masculine potency. We might compare the emanations brought forth from fallow, uncultivated and implicitly feminised land: the weeds, water and subsidences, and the leakages of odour and other substances from houses 'impregnated with disease',[14] with the rich, masculinised outpourings of industrial processes whose saturated, productive capability makes containment impossible.

To pursue such a gender reading might suggest that the courts place greater value on the inevitable byproducts of industrial usage than on the worthless, uncontainable emissions of the untamed, natural estate. At certain periods, undoubtedly, British courts have proved to be tolerant of industrial nuisances (see, e.g., Brenner 1973), particularly where the only harm was to the amenity values of poor or working-class neighbourhoods. However, at other times, over-productive property has been required to rein in or limit its potency (though here the courts tend to distinguish 'normal' industrial emissions from leakages, and both may be distinguished from unnatural and dangerous uses). This 'reining in', through the implicitly agreed compact of a mutually bounded freedom, exists both to protect the productivity of other masculinised industrial spaces, as well as to safeguard the amenity of property owners in their passively feminised domestic or rural idylls. At the same time, courts in England have proved to be reluctant to recognise the interests of the 'overly' delicate plaintiff, process or space, seeking to extend *her* freedom through the imperatives generated by her hyper-sensitivity.[15] Frequently, English courts refer to the need for private nuisance claimants to accept some level of interference: a bounded freedom read through the prism of a hardy, masculine give and take.

The third way in which nuisance refracts relations of inequality concerns the power of institutional structures to shape what gets defined as an impediment. One example of this process is evident in the faculty–student case outlined earlier. However, because this represents a quasi-fiduciary relationship rooted in the vulnerability and expectations of the student and the power of the academic, it provides a less clear example of the ways in which being and causing a nuisance are mapped onto institutional hierarchies. Family relations provide another instance, although here too kinship obligations mitigate claims that the subordinate other is a nuisance. A different instance is that of relations within a company or firm. Employers are likely to read employee conduct as a nuisance to the extent that it impedes – or threatens to impede – efficiency and profitability, for instance, as a result of industrial protests in pursuit of better working conditions or through the establishment of a trade union (see Vorspan 1998). At other times, employers associate employee nuisance externally in those disruptions to labour caused by sick family members, personal problems or chronic transport breakdown.

However, nuisance claims within hierarchical, institutional structures do not just come from the top. Nuisance's terms also reflect, and offer ways of containing, the power of senior management. This carries legitimacy within an institutional context where the rights and authority of superordinates are delimited; however, it also enables workers to deploy nuisance's terms as a means of resisting and repudiating the legitimacy of senior demands.

Reading managerial conduct as impeding occupational autonomy and entitlement can be seen in two contexts. The first concerns excessive surveillance, for instance managers who are always looking 'over the shoulder' of staff or constantly demanding reports. Here, in contrast to the nuisance of subordinates which tends to be associated with chaotic, unpredictable behaviour, superordinate nuisance operates through a spectacle of overly enforced control and the withdrawal of discretion from below. In other words, in a context where 'good' management works through trust and individual self-governance, and takes place 'at a distance', shortening the thread may be read as a nuisance.

Instructions are also interpreted as a nuisance when they are perceived as beyond the scope of the job. What counts as such varies according to a range of factors, including that of gender. Women in clerical and secretarial jobs have traditionally been asked to perform duties, such as buying presents, making non-work related appointments, serving tea and doing the washing up, that would be seen as inappropriate to ask of male employees. While the gendered character of occupational flexibility continues, particularly for female administrative workers, some assumptions of scope have been effectively contested and have changed. One area in which nuisance claims refract evolving perceptions of appropriate workplace conduct is in relation to sexual harassment by a supervisor. In many traditional female occupations, a narrative of sexual availability, oriented towards managers or clients, functioned (at least as far as the employer was concerned) as an integral aspect of the job (e.g. Adkins 1994). In other cases where women directly assisted male professionals, such as dentistry, the assistant functioned as surrogate wife (Adams 2000). Sexual overtures and gendered patronage could there be read as everyday workplace forms of spousal intimacy. Male flirtation, bantering and orders were to be accommodated as constitutive, rather than an impediment, to the performance of the job. A central aspect of feminist political activity, consequently, has been to rewrite the scope of women's work so that sexual interactions are placed outside, neither part of the job nor integral to a pleasant working environment.

Being a nuisance

So far, I have focused on the politics of harm, and in particular nuisance, from one primary angle: the ways in which, far from being socially neutral or benign, harm refracts and sustains dominant social norms, and works to consolidate social dynamics and inequalities. But claims of harm do not simply reflect pre-existing distributions of power; they remain far more dynamic. In the evolution of what counts as harm dominant constituencies also confront limitations in their capacity to act against those with less power. In the final part of the chapter I explore this potential further. Here, however, I want to consider a somewhat different issue, namely, what subordinate and resisting forces can do in the space defined by powerful constituencies as harmful. In particular, what purchase does 'being' a nuisance have within a counter-normative politics?

Unsurprisingly perhaps, nuisance has not proved to be a popular paradigm for political movements contemplating effective strategies of mobilisation. In the drive to win support and to persuade outsiders, being dismissed as a nuisance appears both counterproductive and demeaning. It is therefore a label of, at best, limited appeal. At the same time, given the application of nuisance to conduct and people who are out of place, distracting attention, obstructing social processes or causing sensory offence, its contours establish an inevitable space for a range of political activities. But how productive is this political space? And what limitations does it incorporate?

Being in the 'wrong' place is a core element of movements that challenge the allocation of people and conduct to different and unevenly resourced spaces. While activists may not seek or wish to be defined as a nuisance, their 'inappropriate' presence – resisting their assigned placing – may have this effect. Crossing zones of conduct and identity – witnessed most vividly in the civil rights movement, in the struggles against apartheid, and in feminist protests over male places of exclusion – has something in common, at least at first glance, with the transgressive politics valorised by sex radicals. Transgression, with its breaching of norms of convention, has been readily deployed in struggles over norms of gender and sexuality, as witnessed in the turn of the century interest in transgender practices (see chapter 4), and in the longer cultural stylisation of butch–femme.

Yet while performative transgressions necessarily involve repetition, and to this extent are sensitive to the accusation of being offensive or a nuisance, the breaching of social conventions and appropriation of the stigma of social taboo are often part of, or wrought to generate, normalisation. Crossings which invoke being in an improper body, clothes, jobs, buildings or spaces are often intended to be only temporarily out of place. In other words, they seek to normalise their innovativeness: to become unproblematically accepted. Through repetition what once seemed unacceptable, unimaginable, ludicrous, even offensive, becomes mundane or banal.

A second kind of transgression is of a different nature; it concerns the deployment of symbolic or theatrical modes of activism that use the out of place to gain attention. Theatrical protests such as kiss-ins and wed-ins have been widely used in gay and queer activism as have 'die-ins' by AIDS activists to protest at the silencing and erasure of people living with AIDS as well as the costs of medical treatment (Brown 1997: 64). The action of falling to the floor at an appointed time proved a powerful, evocative form of protest. It generated media attention and helped to consolidate political communities as well as demanding people's attention through its transgression of norms of proper place.

Transgressive forms of protest privilege the one-off – the unexpected, apparently spontaneous, and often short theatrical gesture (see also Brown 1997: 60–1). However, their effectiveness as a means of communicating anger or revealing discrimination is limited to the extent that the action is read as one of nuisance. While acts seen as a nuisance, by their very nature, are never fully anticipated, they also do not surprise. They are the familiar, recognisable action or crossing witnessed before. Nuisances may demand a deep breath, expletive, or question of what now, but they always announce that resolution is possible (even if it involves action by others, such as the courts). By focusing people's minds on their own feelings of stress, irritation, anger or loss, and by reducing that which they confront to an irritant or obstacle, nuisances deflect attention away from the epistemological or normative challenge posed by the spectacle confronted.

My third kind of activism does not shy away from being described as a nuisance. Indeed, it *relies* on being seen as a nuisance for its effectiveness, unlike the transgressive forms of protest described above. Being an obstacle offers a productive means of opposing processes that depend on forward planning, non-interference and efficiency.[16] The effectiveness of private-sector strikes or 'go slows', for instance, lies in their impact on the economic viability of affected companies. Protests and the physical occupation of land, particularly if they prove tenacious and expensive to dislodge, can undermine the profitability of planned developments, such as road expansion, mining and new buildings. In addition, through the publicised willingness of people to place their bodies in the way, occupation can challenge the legitimacy and sound common-sense of new economic developments. Anti-road activist Phil Mcleish describes the effect of being an obstacle in relation to anti-road protests, 'This was politics of provocation, winding up the state and then watching and laughing as the state makes a fool of itself... Forcing the police to spend over £2 million playing games...' (Welsh and McLeish 1996: 39).

Blockages can be insinuated into other systems. This kind of tactic works best where systems depend on regulated use: for instance, banks' reliance on a limited proportion of people withdrawing their money at any one time. A related example is that of jamming telephone (and other communication) systems through mass simultaneous usage to stop particular transactions or to obstruct specific information from being passed on to authorised clients or enquirers. It can also have a more symbolic effect, and of course its purchase extends beyond progressive or radical causes. Hugh Muir, writing in the British *Guardian*, gives the example of a right-wing radio presenter who 'urged listeners to jam the switchboards at [London's] City Hall in protest at [the mayor] Livingstone's refusal to sanction a victory parade for British troops after the war in Iraq'.[17]

The politically nuanced, tactical deployment of disruption and obstacle-formations as a means of troubling or disrupting the status quo is important, particularly in the way in which it moves beyond arguing for change to mobilising sensory perception and more embodied forms of action. At the same time, nuisance actions may seem a rather limited approach, given their emphasis on obstructing present processes rather than offering alternatives. To this extent they resemble the everyday tactical strategies discussed by scholars such as de Certeau (1984), and Ewick and Silbey (1992). Such tactics work on and within dominant social processes. They seek to disrupt the power of the status quo, to thwart its ability to command and control through feints and dodges, but put little substantively new in its place.

One context in which nuisance has been associated with more generative and transformational projects is where communities, pursuing a way of life based on alternative values, are condemned by outsiders as a nuisance. Such communities may be seen as disrespecting the delineations and spatial classifications of the wider neighbourhood or society by, for instance, failing to stop 'displeasing' smells, sounds and sights from violating pre-existing property boundaries. Yet in many such disputes, whether the nuisance is premeditated and intended may itself be contested. In the case of the Greenham Common Women's Peace Camp, Cresswell (1996: 106–7) suggests that the women were condemned for intentionally smelling. In other words, opponents suggested, campers were consciously using the distastefulness of their bodies to bailiffs and police as a way of discouraging officials from coming too close. This claim was strongly rejected by the women living at the camp, who argued that the conditions in which they lived made conventional standards of hygiene impossible.

The perceived nuisances caused by unwelcome emissions may prove to be an unintended side-effect when prefigurative or utopian forms of living occur in the vicinity of more conventional lifestyles, but the reverse is also evident. In other words, collective endeavours to be an obstacle or a disruption – for instance by impeding new developments such as road building – may, in practice, transcend a purely antagonistic or resistant politics by providing a site within which different ways of living can emerge and develop (e.g., Doherty 1996). We can see this extension from doing to also being in industrial actions, such as the miners' strike in Britain in the mid-1980s, in the international protests against global capitalism that emerged at the turn of the twenty-first century in cities across the world, and in the sporadic, mass non-payment tactics against unpopular taxes. Through their emphasis on disruptive doings as well as sayings, these struggles have, to differing degrees and in differing ways, provided spaces in which other values and social relations become – not always intentionally – installed.

The process of creating and inhabiting non-hegemonic ways of living, and the difficulties of doing so against the gravitational pull of the mainstream, is the subject of the two chapters that follow. However, in the final section of this chapter I want to turn to the capacity of nuisance conceptually to frame social relations and practices in counter-normative ways. I have suggested already that claims of nuisance are subject to dispute, that different practices can be defined as a nuisance. However, I now want to push this argument further in order to explore ways of thinking about nuisance which do not simply substitute one act or identity for another, but seek to reconceptualise the terms of nuisance itself.

Nuisance as a counter-discourse

For the most part, nuisance has received little attention from the intellectual left. While some affirm nuisance as a way of blocking and impeding dominant forces or as a way of expressing the repressed energy of the (usually male) dispossessed, and while others, conversely, support regulating antisocial behaviour within working-class neighbourhoods and extending nuisance laws to cover new forms of pollution, mostly those on the left, and within diversity debates, have not seen nuisance as being of much interest. However, as I have sought to demonstrate in this chapter, nuisance is a highly productive concept. In its dominant, mainstream usage, nuisance renders opaque, while simultaneously consolidating, key social norms and relations. But nuisance's relationship to the status quo is not wholly parasitic: like other injury-based terms, nuisance does not simply reflect the world it inhabits but refracts it in distinctive ways. In other words, nuisance highlights and emphasises certain aspects of the social, while distorting or erasing others.

With their emphasis on the annoyance, irritations and other problems that obstacles can cause – the impediments and distractions that 'waste' time (time that cannot afford to be wasted) – nuisance claims largely work to sustain and reproduce dominant readings of accountability, responsibility, proper(ty), negative freedom, the public/private and kinship. In contrast, nuisance discourse maps far less comfortably on to other norms and practices, such as multiculturalism and discretionary forms of welfare. This is not because these latter are less hegemonic. Rather, it is because nuisance discourse skews the social world it frames in ways that catch at the sharp edges of social diversity and that erase needs without status as entitlements.

Can nuisance discourse function differently? As antisocial behaviour, nuisance can – and has been – identified in a diverse range of social contexts – left-wing as well as right-wing, collectivist as well as individualist. So the commune member who fails to work their shifts or who treats collective goods with disregard can be condemned for being antisocial just like the vandal of private property in liberal society. This elasticity largely comes from the formalistic meaning 'antisocial' bears in designating behaviour that transgresses or diverges from dominant or socially accepted norms.[18] But 'antisocial' also carries an additional meaning: behaviour that disrupts the capacity of people to live in harmonious relations with one another. While this was interpreted rather conservatively by both Conservative and Labour British governments of the 1990s, where sociality was anchored in individual rights, economic freedom and public and domestic order, it can also have – indeed, arguably, leans towards – more solidaristic connections. We can see such connections as integral to the concept of nuisance. For at the very moment that nuisance claims articulate the legitimacy of borders, separation and division, they simultaneously underscore interconnection, permeability and unboundedness: the ways in which our senses come into play as a result of the odours, sights, sounds, tastes, time and bodies of others (see also Nedelsky 1991).

Nuisance discourse's progressive potential is not restricted to its capacity to recognise connection, collective interests and the social. As a legal structure and policy problematic it also has the capacity to refract freedom positively and to embed a positive duty to avoid harm befalling others (Bright 2001). While such a duty has had some purchase within private nuisance law, particularly in relation to landowners' responsibilities, it can also be applied beyond the realm of tort law. For instance, delays in paying out welfare benefits, ignoring the request of someone begging, complicating the process of applying for refugee status, forcing pedestrians to circumnavigate busy roads can all be redefined as a nuisance in that they thwart or impede people's ability to access the resources or environment they need to pursue their interests and goals. How we understand nuisance depends on how we understand both entitlement and responsibility. Nuisance claims are contingent on such understandings being in place. The question is whether oppositional or counter-normative nuisance claims can, in any way, bring *changed* notions of entitlement and responsibility into being.

Before turning to this question, I want to sketch briefly the common ground between more progressive applications of nuisance and its current, dominant usage. This common ground, I want to suggest, underscores the conceptual heart of nuisance as it is currently understood and deployed. Whether nuisance is applied to negative or positive freedom, collective or individual interests, to the practices of dominant or subordinate constituencies, it continues to emphasise impediments, interests, entitlements, and visible beings and doings. Nuisance discourse treats problems as warts, obstacles that stand out in relief against a taken-for-granted social, that can be surgically removed while leaving everything else intact. We can see this with ASBOs introduced in Britain through the Crime and Disorder Act 1998. ASBOs are predicated on the assumption that the problem lies in discrete, spatially manifested acts that are amenable to prohibition. As a result, the detrimental consequences of more subtle, systemic or reproductive processes, including the capacity of such processes to create the very impediments nuisance claims simultaneously and subsequently condemn, become disregarded.

One special quality nuisance discourse possesses, in this respect, is its capacity to replay contested norms and relations as common sense. This enables nuisance claims to operate effectively *within* a community, including within minority or oppositional communities (see chapter 8); however it does not facilitate dialogue *between* different moral constituencies. Social movements may generate loud utterances about the nuisances they encounter, but to the extent that these utterances remain counterintuitive, this is an internal conversation. Attempts to identify nuisances against the terms of dominant values are more likely to generate bewilderment or non-comprehension. This casts doubt on the capacity of nuisance discourse to embed values lacking widespread support. While nuisance claims may help to consolidate the status quo, both within a community as well as within wider society, it is not clear that such claims can help to reformulate terms such as entitlement and responsibility in counter-normative ways.

But this does not make nuisance claims irrelevant to pursuing a radical politics. First, as several progressive and feminist writers have explored, nuisance as a regulatory structure can be used to counter newly recognised harms, such as environmental pollution and sexual harassment (Conaghan 1993, 1996; Conaghan and Mansell 1999; Wightman 1998). Reframing these practices as a nuisance revokes their entitlement, even if it sometimes appears to overestimate the ease with which they can be eliminated. Second, we can identify the productivity of nuisance claims in the uncertainty they generate, as the recipient of the utterance wonders why an accepted practice, such as police officers' interference with begging, is being depicted as an irritant or obstacle. For such a redefinition demands that the listener reconsider the vantage point from which they view the scene as well as the values and assumptions on which such a viewing is premised. Nuisance utterances in oppositional ways may therefore kick-start productive dialogue and debate. The trouble is that this rarely occurs. The emphasis of modern political strategy is to use language that will convince and persuade outsiders, not bemuse them. So, to the extent that terms such as 'harm' and 'nuisance' are deployed in political struggle, the tendency is to deploy them in ways that silence or suppress any reflection on normative differences, through the (implied) imperative of an instinctive unity.

Third, as I have suggested, terms such as 'nuisance' can operate productively within bounded oppositional communities to entrench and normalise counter-hegemonic values. The contingent and contested quality of injury-based terms, their dependence on particular ways of understanding social relations and norms, does not render them redundant. For terms such as 'nuisance' or 'harm' embody important ways in which people understand their relationship to the world around them, and provide a means both of community-building and of value-amplification. In addition, to the extent that such communities impact on dominant lifestyles and practices, the development and solidification of an interior common sense sustains pressures on the status quo. And by pressures I do not simply mean the generation of counter-discourses or political demands.

Rather, I mean the range of ways in which counter-practices affect political authorities, mainstream beliefs and resource allocations.

This quality of prefigurative and utopian communities proved an important element in the aspirations of radical forces of previous decades, aspirations severely dented in more recent years. In the final three chapters of this book, therefore, I revisit the political possibilities posed by 'alternative ways of being' that do not simply advocate new social and normative principles, but, more fundamentally, seek to establish and live them. In doing so, my focus is threefold: how are oppositional social practices sustained? What difficulties do they face? And what possibilities does the creation of prefigurative networks and communities, in particular, offer to a politics intent on countering and re-forming the mainstream?

Notes

1. My focus, in this chapter, is amenity rather than physical damage. There is some discussion within nuisance law about which constitutes the more authentic form of nuisance. My reason for sidelining physical damage is that it deviates from the use of nuisance within everyday speech and is less focused on the repetitive forms of conduct with which I am particularly concerned.
2. Concern about antisocial behaviour on public housing projects and in city centre streets led to a range of policies and initiatives that included, in the United Kingdom, the Anti-Social Behaviour Order, introduced by the Crime and Disorder Act 1998 to restrict the movement of people perceived as causing a nuisance; see generally Hughes (2000), Hunter and Nixon (2001); Winter (2000). However, an ASBO has also been used to ban the public use of racist speech, 'Man banned from saying "Paki" ', *Guardian*, 13 Aug. 2003. See also generally Scottish Executive, 'The use of civil legal remedies for neighbour nuisance in Scotland', http://scotland.gov.uk/cru/kd01/blue/rem-12.htm.
3. *Bradford Corporation* v. *Pickles* [1895] AC 587.
4. The distinctions, however, also depend on context and are thus subject to differing interpretations, as Vorspan (1998) discusses in the context of industrial action. Here, we can see a correlation between the value placed on the attention-demanding activity, and the threshold level required for attention demands to be identified as a nuisance (see Vorpsan 1998).
5. One example, taken from Michael Walzer's (1986: 471) writing, concerns the way in which political leafleting and activity may be seen as a nuisance by shopping mall owners, because it distracts the attention of shoppers away from more commercially profitable activity. However, the response of shoppers to the suggestion that their mental freedom is obstructed will obviously vary.
6. In other cases zoning works to create intensified sites of deprivation and instability.
7. For two recent cases, see *Page Motors Ltd* v. *Epson & Ewell* (1981) 80 LGR 337; *Lipiatt* v. *South Gloucester Council* (1999) 9 *Local Government Review* 6–9.

8. *Lippiatt* v. *South Gloucestershire Council* (1999) 9 *Local Government Review* 6–9.
9. *A-G* v. *Corke* [1933] 1 Ch 89, *per* Bennett J, 94.
10. *A-G* v. *PYA Quarries* [1957] 2 QB 169, *per* Denning LJ, 191.
11. *A-G* v. *Corke* [1933] 1 Ch 89.
12. At the same time, the ability of particular spaces to accommodate and put up with nuisances may be a factor in their representation as cosmopolitan and urban.
13. S. 11 Crime and Disorder Act 1998 also explicitly refers to harm caused by children to persons not of the same household.
14. See *R.* v. *Parlby* (1889) 22 QBD 520 at 525.
15. Judicial lack of sympathy for the overly anxious, even hysterical, feminised claimant is clear (e.g., *Miller* v. *Jackson* [1977] 3 All ER 338). What counts as hypersensitivity has also taken on a new relevance in relation to protecting organic farmers from genetically modified seeds (see Lee and Burrell 2002: 531–2).
16. Less corporeal forms of nuisance, such as graffiti and textual reinscriptions can also obstruct or 'distort', by targeting the communication of politicians, companies or the mass media. Billboard graffiti can obviously be repeatedly removed; however, its capacity to be continuously replaced can make certain forms of commercial, unmediated communication impossible. See also 'How to be an obstacle', a conference held at the Institute of Contemporary Arts, London, 10–11 Nov. 2001.
17. See 'Shock jocks against the mayor', *Guardian*, 11 Aug. 2003.
18. Interestingly, British parliamentary debate and ministerial statements regarding the introduction of Anti-Social Behaviour Orders explicitly stated that they were not intended to be used against 'eccentric activities that are peculiar to individual minority ethnic communities' (Russell, *Hansard*, HC, 1 Feb. 2000, col. 142). According to Alun Michael, Home Office Minister, in his speech to the Targeting Anti-Social Behaviour Conference, Nottingham, 12 June 1997, unusual or non-normative behaviour would not be penalised unless it threatened the safety of the community (http://www.users.globalnet. co.uk/~kings/michael.htm).

7

Oppositional routines: the problem of embedding change

So far, *Challenging Diversity* has attended to the question as to how to identify and undo relations of inequality in the light of the social processes and norms that underpin them. However, while I have explored different strategies and techniques, a central premise of my argument has been the impossibility of pinning change down. So, in chapter 4, I argued for individual equality of power as a political aspiration, to be pursued through dismantling social inequalities. At the same time I suggested that equality of power was an unattainable goal, while its pursuit needed to confront the likelihood of new inequalities arising. This fluidity and lack of certainty does not make change pointless, however. It simply makes it more complex. In this chapter and the one that follows, I address one particular aspect of the change process: the introduction of new sustainable practices and routines. More specifically I ask: what techniques and strategies are needed to create new, embedded social practices? But this question has a twist, since the practices in which I am interested are those at odds with the wider status quo.

For the most part, diversity politics has paid little attention to this issue. More has been said by liberal multicultural writers concerned with the dilemmas posed by assimilation, on the one hand, and social accommodation, on the other. However, while such writing has focused on the problem of survival for minority practices, its emphasis on ethnic differences means that other differences, particularly normative ones, become bracketed. This is the focus of this chapter, which addresses the question of how to create and sustain ways of being and doing that counter, thwart or trouble dominant social relations. There is a tendency in thinking about this issue to assume that new ways of living emerge as a result of ideological or normative commitment. To some extent this is no doubt true. However, the crux of my analysis is the claim that commitment alone may not be enough to sustain counter-normative practices when they come up against the countervailing pressures of the mainstream. This is hardly a novel point. However, it tends to get lost in the cries of hypocrisy that greet people unable to sustain the practices they believe in, whether it is vegetarianism, ethical management, state education and healthcare, non-monogamy, non-property ownership or anything else.

To explore these issues more fully, the chapter is divided into two main sections. The first considers in more detail what value new routines offer from the perspective of a radical politics. It then proceeds to construct a way of thinking about routinised practice, drawing on the metaphor of social pathways. The second part of the chapter examines the conditions necessary for new counter-normative pathways to be established. My focus is the role played by the social environment. To explore this in more detail, the chapter considers two attempts to create new trajectories of action, in the first case without being able to reconfigure the environment, in the second case by introducing environmental changes as well.

Creating sustainable practices

The construction of new routinised pathways, from recycling to lesbian motherhood, contributes to the pursuit of a radical politics in several respects. It maintains pressure on incompatible aspects of the status quo, a pressure that comes from the repetitive encounters between 'awkward' forms of doing and the mainstream. It demonstrates other modes of living in ways that convince through practice and illustration rather than through argument alone, and, for those who pursue them, provides a way of experimenting with different forms of existence. To the extent that new practices become routinised or taken for granted, they may prove harder for opponents to attack and easier for proponents to pursue. Once traditions, continuity and history are possible, pathways engender roots, stability and perspective. They also promise 'rightness' and propriety – of being in place – things frequently lacking, even for believers, when oppositional pathways appear transient and ephemeral.

In arguing for the value of routinised or enduring non-dominant practices, it is worth briefly considering, however, the counter-claim that both mindless reiteration and durability are antithetical to a radical politics. This claim accepts the possibility of oppositional practices becoming sustainable, for instance by appropriating the authority of law, but argues that such a development should be denounced, not valorised, since it leads to new forms of hierarchy, discipline and exclusion. Linked to this critique is another, opposed to the stability of counter-cultural practices on the grounds that routines and repetition eliminate the benefits of spontaneity, while constraining freedom and self-awareness.

In adopting a position which asserts the value and importance of routines and reiterative practices, my analysis also diverges from those who see repetition and routines as antithetical to moral conduct (e.g., Bauman 1993). Rather, I want to suggest, the moral conduct of some depends on the conditions of possibility being created and maintained by others – a process which requires norms, standards and met expectations, even as it also requires resistance, transgression and creativity. Likewise, liberation is not the installation of fully reflexive choice, for the complete absence of routine would bring forth its own tyranny. If every gesture, movement or chain of actions required full consideration, agency would grind to a halt. For some actions to be the product of reflection, others need to remain below the deliberative threshold (see also Campbell 1996). This may be because they are taken for granted truths or beliefs; it may be because they are too obviously 'rational' to require conscious reflection. Sub-deliberative action can also function as a form of bodily knowing or 'hexis' (Bourdieu 1977: 93, 94; Jenkins 1992: 75; see also de Certeau 1984: 93), where 'strings of "moves"' (Bourdieu 1992: 62) acquire a corporeal common sense.

Social pathways

Are there forms of repetition that do not constitute a simple imitation, reproduction, and, hence, consolidation of the law . . . Which possibilities of doing gender repeat and displace through hyperbole, dissonance, internal confusion, and proliferation the very constructs by which they are mobilized?
(Butler 1990: 31)

If establishing new routines are an important aspect of social change, how do they emerge? Addressing this question requires us to consider not just how new practices come into existence but how they are able to survive for long enough to become routinised. While a number of scholars have explored the ways in which dominant social practices reproduce themselves over time, less has been said about the staying power of oppositional or minority practices. Unlike those of the mainstream, these tend not to be located within a congruous social environment; as a result, they confront ideologies, resource allocations, institutional practices and social norms that undermine rather than underpin. What chance do they then have for survival?

To explore this issue, I want to consider the creation of new strings of practices drawing on the metaphor of social pathways. This metaphor has been widely used to describe a range of processes, from an individual's life trajectory to routes into and through higher education, and other institutionalised social experiences such as offending (Bickley and Beech 2002; Kapp 2000) and homelessness (Butler and Weatherley 1995; Tessler et al. 2001; Weitzman et al. 1990). Used in this way, pathways connote the complex trajectories or routes that take people in particular directions. Weitzman and colleagues (1990: 133), for instance, in exploring the pathway from housing to homelessness considered the factors – the milestones, losses, problems and precipitants – that placed people on this trajectory; they also explored the varied gradient of the pathway, how for some social groupings it was short and abrupt, for others more gradual. Kapp's (2000: 73) study of the pathways of young offenders adopts a broadly similar approach: 'There seems to be different impressions of the pathway from the juvenile system to the adult system depending on the original reason for entry.' What 'pathway' here emphasises is a trajectory of environmentally shaped action over time – in other words, conduct has roots in prior events and socially configured 'choices'. It also suggests that while 'entry' actions or conditions may lead to a particular 'end-point', it is not inevitable. People confront different junctures, where the routes chosen may take them further into offending behaviour and homelessness or out of it.

'Pathways' has proved to be a popular term. For policy-makers it provides a useful way of framing the need for a flexible and proactive response. If the routes into homelessness, and their association with particular social indices, can be identified, it may be possible to design policies that reroute particular pathways away. My interest, however, in the pathways metaphor is from a somewhat different angle. The usefulness of the term lies, for my purposes, in two other respects. First, 'pathways' emphasises the diachronic connections that people create, articulating different activities across time and space. Pathways therefore complements the concept of practice which tends to emphasise less personalised synchronic connections within a field.

The second contribution 'pathways' makes is in its association with land. At this level, the metaphor works effectively to invoke the ways in which the environment sustains and helps to shape the pathways that cross it. The path metaphor also evokes the processes by which particular routes and routines can become secured and deepened (see also Ingold 1993: 167). The figurative retreading of the same path, caused by the repeated journey of one traveller or the common trajectory of several, creates erosion, establishing the path further and making it harder to miss. Paths lead towards and away from places, although some may just exist for the enjoyment of the walk itself or for the features along its route. Paths intersect, creating opportunities and demanding choices, risks and gambles, as the direction may not always be clear or knowable. In disuse, paths become overgrown and can vanish from sight.[1]

One example of the evocative metaphorical value of pathways can be found in the writing of the transgender author, Riki Wilchins, as she movingly portrays the relationship between social pathways and her individual, non-conformist trajectory.

> I have spent my life exploring the geography of this place [as a lesbian, transexual, femme, incest survivor, 'addict'], mastering unfamiliar terrain and alien customs, wandering regions as fresh, as uncharted, as inexplicable to me as private visions. I have surveyed its pathways, as ignorant ... as any first-time explorer, and finally found myself at day's end – lost, alone, bewildered, and afraid. With time, my tracks have intersected and converged, crisscrossed again and again, until at last they have woven their own pattern ... (Wilchins 1997: 173–4)

Yet despite the capacity of the pathways metaphor to illuminate diachronic relations between events, and to highlight the different ways in which trajectories can become, through repetition, socially embedded, it also has limitations. I want briefly to mention three and then explore one way of responding to these constraints. The first relates to the question of agency. Writers using the pathways metaphor have adopted different approaches to this. For instance, Dunn and Hayes (1999: 399) adopt a relatively structuralist approach in suggesting that 'one of the most important research needs in health inequalities scholarship is to better elucidate those pathways by which differences in socioeconomic status manifest in everyday life, and produce . . . the systematic social gradient in health observed in all industrialized countries'. In contrast, Butler and Weatherley (1995: 9), in talking about homelessness, write, 'Each woman's story is unique, and there was no direct or simple pathway to their homelessness.'

Part of the problem with the way in which the pathways metaphor is generally used is its emphasis on pre-existing routes that go beyond and pre-date individual action. While scholarship on 'real' walking recognises the possibilities for 'improvisational trails', 'wandering, lounging, and "hanging around"' (Edensor 1998: 106, 99), this is less evident in social policy's use of the pathways metaphor. Even when individuals are seen as making choices, or exercising agency, this is usually in a context where the effects of making particular choices are predetermined. To the extent that the metaphor is used to understand why people offend, become sick or find themselves homeless, it offers a teleological framework in which particular decisions or social factors are understood through a process of tracing paths back from their 'end point' to flag up key linkages and junctures.

Two other problems are identified by de Certeau (1984: 35) in his discussion of the term 'trajectory'. Although trajectory and pathways are not identical, the issues he raises have applicability here too. De Certeau asks whether trajectories – or in our case, pathways – emphasise diachronic (or temporal) connections at the expense of synchronic (spatial or functional) ones? Do paths reduce 'time and movement . . . to a line that can be seized as a whole by the eye and read in a single moment' (de Certeau 1984: 35)? De Certeau also asks whether such figurative terms impose a 'reversible sign . . . for a practice indissociable from particular moments and "opportunities", and thus irreversible' (ibid.)? In other words, does the metaphor of pathways ignore the specificity of social conditions at a given moment, in the suggestion that common ground exists not only between those travelling the path in the same direction, but also among those travelling 'backwards' along the same route?

De Certeau (ibid.) goes on to argue that the 'distinction between *strategies* and *tactics* appears to provide a more adequate initial schema'. However, I want to hold on to the language of pathways, and deal with de Certeau's instructive criticisms in two ways, first, by using the mapping techniques of scale to explore the distinctively *social* character of the pathways I am here concerned with. Second, against de Certeau's concern that pathways form 'flattened' cartographic representations, I want to distinguish de jure pathways that are imagined, represented and constructed usually from 'above', from de facto pathways that become embedded, but may also remain emergent, through the process of being walked and woven. This distinction between de jure and de facto pathways also provides a way of addressing the question of agency identified above.

The metaphor of scale makes it possible to differentiate between detailed individual paths, with all their variations and idiosyncrasies, organisational and corporate paths, and what I want to focus on here: social routes that become abstracted from, while also shaping, the overlapping tracings of individual trajectories. What counts as each will vary depending on the context. So individual pathways in some cases may refer to individual households, couples or organisations. In other instances, it is the diverging and intersecting pathways *within* these entities that are subject to scrutiny. *Social* pathways function simultaneously at different temporal scales; these range from the cartographic large-scale, micro-trajectories of daily workplace or household conduct, to the annual routines of traditional and festive life, and the smaller-scale passage of our lifespan, where iteration comes through commonality with others rather than individual repetition.

Represented cartographically, small-scale social pathways lose the nuances and sometimes the sequencing of individual paths in cases where they become opened up to a reverse flow of movement. Boaventura de Sousa Santos (1987) argues that each map scale reveals certain characteristics or phenomena while distorting or obscuring others. Here, in respect to social pathways, what gets highlighted in broad brushstrokes is the relative significance and direction of different trajectories as well as the ways in which practices interlink. In other words, the cartographic scale of social, rather than individual, pathways reveals well-established trajectories and well-used intersections. It also reveals major contours: the hills, valleys and settings navigated, the places where the route breaks down, disappears or stops short. However, what cannot be seen, from this height, are one-off journeys or the nuanced individual tracings that criss-cross more established or well-traversed paths.

This kind of 'visualisation technique' makes it possible to respond to the criticism that the metaphor of pathways ignores social variation, relationality and unfinished individual trajectories. The pathways metaphor opens itself up to criticism in this respect since it not only assumes that action is socially structured but that the actual patterning of action, as engaged in by individuals and institutions over time, has a common character. But the extent to which and ways in which it does demonstrate commonality, I want to suggest, is an empirical point. While we may expect that looking from a microlight aircraft we will see broad, intersecting paths cutting a swathe through the landscape, whether or not we see them requires that we do in fact 'look'. For where we expect to see junctures there may in fact be hermetic loops; instead of one path there may be several, of different dimensions and trajectories, or even – despite our shallow height – none at all.

De facto and de jure pathways

To explore more clearly the relationship between individual and social paths, I want to focus on the way in which new pathways emerge. Borrowing loosely from the terminology of international law, I want to distinguish between two primary, although complexly intertwined, processes of path-building: de facto – through usage – and de jure – through governance techniques. The distinction between them – lost in highly voluntaristic accounts where they collapse into the former, or highly deterministic or structuralist accounts where only de jure processes exist – is central to the question of creating sustainable counter-normative trajectories.

While some de facto paths emerge as the conscious products of creativity, resistance or force, others are less intentional – generated by the 'logic' of their social environment. The term 'logic' is an important one for my discussion. I do not use it to identify a specific form of thought, but rather to highlight a distinctive and important way in which the social environment *steers* practice. Social logics shape preferences, interests and identity; however I am also concerned with the ways in which the environment shapes conduct *despite* the countervailing interests and preferences of the people concerned.

Intelligible through the lens of self-interest, concern for others or normative organising principles, social logics work to constitute certain actions as profitable, efficient, safe, pleasing, desirable, normal or rational. Others, in contrast, are read or experienced as ridiculous, irrational, embarrassing or costly. These two processes frequently overlap so that, for instance, conduct that is normatively authorised may, at the same time, prove exhausting or emotionally costly. Struggles over the relative good sense of particular pathways or routines can be seen in the tensions and conflicts over car use, 'unprotected' sex, vegetarianism and full-time motherhood. These struggles may involve one side highlighting costs previously hidden or reading activities in the light of new and different norms (see also Benton 1994). They may also emerge as a result of changes within the social environment that cause new logics to be generated.

In exploring the relationship between the environment and de facto pathways, the role played by social inequality is an important theme. In a socially differentiated society, pathways are not equally accessible or equally navigable by all. While prevailing conceptions of equality within liberal, Western societies focus on opening up the terms and conditions of formal access, environmental factors shape, in unequal ways, the actual pathways pursued by different social constituencies. Let me clarify what I mean by this, perhaps most easily by outlining what I am not saying. First, I am not suggesting that the environment invariably constructs distinct and separate pathways. While exclusive or 'gated' trajectories may exist, such as heterosexual marriage or age-related schooling, these are nevertheless penetrated and structured by other social inequalities, particularly of class, race and ethnicity (see also Tessler et al. 2001). In the West, pathways tend to be produced and woven out of a bricolage of different social locations and relations. But this does not mean a cascade of finely tuned combinations – the pathway of white, middle-class, disabled women, for instance – although in some contexts such pathways may be found. It also does not mean that de facto (as opposed to de jure) pathways can be read from the modes of power that operate. How the environment structures social action cannot be simply arrived at by adding the disciplinary, coercive, ideological and resource-based technologies of power in operation. I discuss the complexity of the environment's impact further below. However, its relationship to agency and social inequality is evocatively evident in Riki Wilchins's (1997: 36) description of the pressures borne by transexuals and 'genderqueers', where the pathways pursued are crafted out of a grammar of self-expression, safety and wellbeing.

In order to survive, to avoid the bashings, the job discrimination, and the street-corner humiliations . . . [s]he must know how others see her so she can know how to see herself . . . She will gradually learn how she looks and what her body means. She will carry this knowledge around, producing it on demand like pocket ID when she enters a subway car, applies for a job, approaches the police for directions, uses a women's room, or walks alone at night past a knot of men.

While pathways can be created and secured through usage, new paths are also established de jure, that is through governance techniques. In this case, paths are mapped and constructed for others. State bodies are particularly prominent in building paths de jure, combining their power as law- and policy-maker, purchaser, service provider and employer. However, community bodies may also seek to establish de jure pathways for members, as I discuss in chapter 8. Frequently, new paths are established through a combination of de jure and de facto means. The eruv, for instance, that I discussed in chapter 2, is a technique of Jewish law that, by establishing new expanded limits to permissible paths, enabled different de facto Sabbath routes to be forged. Although many social pathways are established through a combination of de facto and de jure means, pathways dependent on the latter are often saturated with tensions. This can be seen in the case of new controversial legislation, where the tentative, compromise pathways, mapped by government spokespeople and others, are scarcely recognised in the haphazard 'new build' brought into being by street-level and other actors. As a result, legal actions, circulars, and amending or clarifying legislation may be deployed to bring pathways closer to the original plans. Such a response aims to close off the possibilities for front-line actors to maintain de facto routes based on pre-existing rationalities (see, e.g., Cooper 1995b, 1998a). At the same time, government action frequently fails to compel compliance, as I discuss further below.

The need for additional action to complete the links between pathway mapping and effective construction raises questions of durability. De jure paths may be sustained through government labour and regular clearance operations to eliminate debris. But if de jure pathways receive little use, and especially where they lack legitimacy, they are likely to fall into disuse, once government attention has waned (see also Barley and Tolbert 1997: 112; Oliver 1992: 570). In contrast, well embedded de facto paths may endure despite government disfavour. Public bodies may create new maps from which the paths have vanished without trace, but in the thicket of everyday practice, unfavoured or even illegal trails – officially suppressed religious rituals or national identifications, for instance – may continue to be pursued.

The capacity of utilised paths to outlive their conditions of production is important. It suggests that, once embedded, paths can endure, at least for a while, despite their incompatibility with new prevailing norms, resources and wider social or government processes. This does not mean that such 'archaic' paths fail to adapt or evolve (although they may), but simply that some continuity of form, substance or motivation remains. This holds out hope in cases where reactionary changes to the social environment are imposed on top of more progressive traditions. Where pre-existing progressive paths confront a changed or renewed landscape, they provide a means, however temporary, through which historical traces of different forms of social organising can be glimpsed. To the extent they can endure, they also maintain pressure on dominant forces. Bourdieu (1992: 62) makes a similar point when he suggests:

The tendency of groups to persist in their ways, due *inter alia* to the fact that they are composed of individuals with durable dispositions that can outlive the economic and social conditions in which they were produced, can be the source of misadaption as well as adaption, revolt as well as resignation.

However, over time, the ability of oppositional paths to sustain themselves against the regulatory and normative pressures of the mainstream may prove impossible. To the extent that they impel conduct against the mainstream while failing to change it, new routines confront the harsh, relentless pressures of an unconducive environment. The effects of living against the mainstream can be seen in the experience of same-sex couples, particularly their turn in the 1990s towards seeking spousal recognition (see chapter 5). Alongside more conventional campaigns intent on broadening the legal definition of spouse and family, marriage – as a social rather than as a specifically legal performance – became a site of lesbian and gay 'mass trespass' through non-state (though sometimes clerically recognised) commitment ceremonies.

Attempts to make legal categories – here spouse and marriage – more widely available constitute a recognisable aspect of the struggle for equality. However, the emergence of spousal recognition as a key political aspiration also reveals, as I discussed in chapter 5, less progressive tendencies. There are many contributing and interlocking reasons why spousal recognition emerged as a key demand within lesbian and gay movements internationally. Nevertheless, one factor was the difficulty of sustaining counter-normative pathways, organised around non-possessive, non-coupled relationships, when norms, policies (e.g., family, housing and taxation), cultural fantasies and economic arrangements within the wider environment continued to steer people towards non-transient (if not permanent) spousal arrangements.

Environmental conditions

What has emerged from my discussion so far as central is the importance of the relationship between pathways and their environment to achieving durable social change. Whether this is a relationship based on compatibility and convergence or divergence and tension will affect not only whether pathways emerge, but, importantly, whether they survive or fade over time. Organisational analysis, and the debates that have taken place over new institutional theory, highlight many of the environmental factors that shape and structure action (e.g., DiMaggio and Powell 1991; March and Olsen 1983). While the new institutionalism has largely focused on organisational conduct, in my view the analyses also have applicability to other situations. In the chapter that follows I explore the significance of the social environment to the success and survival of intentional and prefigurative community practices. There, I focus in particular on the issue of permeable boundaries, both in terms of how the 'outside' can undermine new community pathways, as well as how permeability can enable members to impact on wider society. Here, however, my focus is on the organisational process of establishing new counter-hegemonic pathways that seek to promote and inhabit new norms and social relations – progressive, but also right-wing. My examples centre on attempts by local and central government to install new pathways de jure and the ways in which these attempts – thanks to the environment within which they emerged – produced other de facto routes. My discussion of organisational pathways thus provides a somewhat different 'take' from the chapter's focus so far on broader social pathways. However, I have chosen this orientation for two reasons: first, so that I can explore the tensions that emerge in the course of pathway mapping, construction and endurance; and second, because the conduct of institutional actors in crafting formal and informal (including surreptitious) pathways is a crucial, though oft-neglected, factor in creating, but also sustaining, social change.

In taking environmental factors as my focus, I want to avoid identifying the environment as the sole or overriding influence on action. Nevertheless, my choice of focus is for two reasons. First, I want to provide a counterbalance to commonsense policy thinking which sees change, almost exclusively, as the product either of top-down instruction – the introduction of correct procedures and new laws – or from training programmes that strategically reshape actors' beliefs and values. While these approaches may promote desired change, they may also simply produce a fruitless Mobius strip where legislation, procedures and training, in their failure to deliver intended results, generate instead an intensification of the same governance techniques. My second reason for focusing on environmental considerations is to centre those factors which are less amenable to immediate control by the organisation or individual in question. Thus I am interested in the divergence that exists between political commitment and 'toned-up' procedures – the new value-driven pathways organisations map – on the one hand, and the actual practices formed 'on the ground'.

Yet if a key factor in this analysis is the environment, we need, before going further, a sharper conception of what it entails. The focus within much institutional theory on relations between organisations has led some researchers to depict the environment as an exterior space in which organisations are located (e.g., Brunsson and Olsen 1993: 124–5). Whittington (1992: 699, 701), in contrast, has suggested that the environment should be seen as more than the external context. But if the environment is not only constitutive of the inside, but also present within interior organisational spaces, does this mean that the environment is everywhere and everything?

In depicting the environment as something more than the fixed landscape against which social and organisational life are constructed – as including cultural norms and relationships as well as resources, opportunities and penalties – I do not want to offer so expansive a definition as to render the environment analytically useless. Rather, I want to use the environment to identify the (often contradictory) complex of discourses, rules, technologies, resource allocations, norms, (networked) relations and practices which structure, *without determining*, the pathways taken. Since the environment depends on the *manner* in which action is shaped, what constitutes the environment will vary according to particular subject positions; it will also vary according to whether the subject is an institution, collective or individual. This kind of subject-centred approach has been criticised within organisational theory for offering too narrow a focus. However, it is useful in highlighting the factors that affect individual conduct, and the extent to which agency is structured or alternatively withdrawn in particular contexts. We can see the distinction in the relatively different positions of a local authority and front-line service worker in relation to local government finance. While financial regimes may seem to function environmentally from the standpoint of councils faced with an array of income or financial options, for many street-level employees budgetary reductions will appear to cut down rather than structure their discretion.

Yet, as even this minor example reveals, the distinction between the structuring environment and the formal, official, often hierarchical, connections mapping people, processes and resources is not clear-cut. Aside from the fact that laws can provide a framework for discretion or identify principles which must be considered in practice, many officially constructed procedures and hierarchical systems of authority are 'environmentalised' in the exercise of actors' agency. In other words, actors mobilise choice, counter-norms, and discretion to personally craft and modify orders intended to be operationalised without variation. This may involve reading new laws through pre-existing discourses, as I discuss further below. It may also entail ignoring new legal requirements (Smart 1995: 156–7), such as where actors attempt to retain discretion over relations, processes and decisions from which it has been officially removed (see also Cooper 1998a: ch. 5; Ellis et al. 1999). At the same time I do not want to suggest that local actors' capacity to 'environmentalise' new laws against the grain is complete (see also Goddard 2002). Not only are some legislative changes more prone to being environmentalised than others, but the process of environmentalising itself is frequently a site of contestation,[2] a site that becomes intensified in contexts where superordinate actors seek to eliminate the build-up of 'counterproductive' discretion.

I now want to turn to two attempts to create new, sustainable, counter-normative pathways. The examples concern local and central government in Britain in the 1980s. I have chosen this period because it illuminates the struggle to develop and inhabit new de jure pathways in the context of considerable political antagonism. My first example relates to attempts to develop pro-gay policies; the second concerns more conservative developments. This latter example highlights the way in which new pathways may seek to challenge the status quo from a rightward direction as well as from the left. In exploring these two instances of pathbuilding my aim is threefold: to consider the relationship between the pathways that governing bodies seek to create and the social environment they confront; to flag up the unpredictable character of the de facto pathways that actually emerge; and to examine briefly the relationship between normative beliefs and pathway practice.

Confronting an inhospitable environment

In the early and mid-1980s the British new urban left (NUL) took control of many local government authorities. Their agenda was an ambitious one, intent on transforming the social pathways of officers, local inhabitants, even people geographically distant, in order to put pressure on the reigning Conservative government and to presage a future, socialist nation-state. Measures and policies introduced included environmental, anti-racist and anti-sexist initiatives, the establishment of new democratic fora, and community and economic developments (Ball and Solomos 1990; Ben-Tovim et al. 1986; Boddy and Fudge 1984; Gyford 1985; Halford 1992; Lansley et al. 1989; Loughlin 1996). While many of the areas to be developed attracted media interest and controversy, the initiatives that were to generate the most opposition and debate were in the equal opportunities field. In many respects these initiatives were grounded in accepted liberal, multicultural thinking. However, introduced by high-profile left-wing councils, they came to function as the front line in the 'culture wars' of the 1980s, symbols of the struggle to install a new hegemony.

By the early 1990s many of these initiatives had faded. Reasons for this are complex, but include the tenaciousness of Conservative national rule, local government's financial retrenchment and forced privatisation, and general demoralisation on the left. However, one cluster of factors to jeopardise sustainability concerned the way in which more controversial new policies were introduced. Because many failed to be properly embedded and so to become routinised, they were able to be picked apart easily once support declined. Indeed, in some cases, no overt dismantling proved necessary as new de jure pathways – increasingly ill-maintained – fell into disuse.

Within organisational and institutional theory, extensive debate has taken place over the relative significance of different environmental conditions, operating at different geographical levels, to structuring organisational activity: from technical factors, such as customers, competitors, raw materials and scientific know-how, to cultural and normative influences (see generally Aldrich and Mindlin 1978; Greenwood and Hinings 1996; Grendstad and Selle 1995; Haage 1978; Kraatz and Zajac 1996; Lawrence and Lorsch 1967; Lowndes, 1996; Scott 1983). The interplay of these different environmental factors, and their effect on the development of new, enduring pathways, can be seen in relation to the attempts of a small handful of NUL councils during the 1980s to promote lesbian and gay equality (Cooper 1994; Lansley et al. 1989; Tobin 1990). These attempts were not governed by a radical revisioning of equality such as I discussed in chapter 4. Rather, they sought to extend prevailing principles of equality, based on opportunity, non-discrimination and empowerment to a 'new', controversial identity in ways that recognised sexuality as an indicium of inequality rather than merely as a form of conduct.[3]

Amidst considerable media attention (and consternation), new pathways for the delivery of sexual equality were mapped out in local Labour Party manifestos, council documents and speeches; the ground was prepared by the appointment of new staff, creating new committees and mini-departments (units), and through the introduction of policy statements of intent. The range of potential new pathways embraced by the initiative was impressive. They included new policy-making and implementation processes, revised conduct by front-line workers, pathways of 'freedom' for lesbian and gay residents, and the pursuit of non-discriminatory pathways by others (including local companies, school students and community groups). Yet, despite their rough construction, and the ongoing rerouting and path maintenance performed by policy proponents, most particularly specialist lesbian and gay staff, the pathways carved out for the delivery of sexual equality never became fully operationalised, embedded or used – a problem repeated, though to a lesser degree, in the late 1990s, when a new generation of lesbian and gay equality work emerged (see Cooper and Monro 2003). The role played by certain environmental factors in obstructing the development of new pathways during the late 1980s is well known: the constraining financial, legal and discursive regime mobilised by central government; the heavy onslaught of the national and local press – heedless of municipal attempts to mollify; and the seasonal effects, particularly on marginal local councils, of having to contest regular elections while safeguarding Labour's electoral chances on the national stage. However, I want to flag up two other environmental elements; these highlight the power of institutional concerns as well as the complex, entwined relationship of the interior and exterior environment (see also Lowndes 1997).

The first concerns the way in which institutional norms, constituting a logic of appropriateness (Lowndes 1996; March and Olsen 1989) within the wider local government environment, structured the bounds of acceptable lesbian and gay policy development, on occasions despite the more radical beliefs of many municipal actors involved. As a result, certain potential pathways remained unmapped – particularly those entailing a critique of heterosexuality or the promotion of lesbianism as a political choice. Others, grounded in a pluri-cultural sense of equality, were mapped but inadequately constructed, or built but inadequately used. Local government norms, signalling the importance of 'non-ideological', appropriate conduct, coincided with municipal and wider perceptions of sexuality as private, sensitive and controversial. As a result, the mass media and other critics were able to portray pathways intended to deliver liberal equality, such as 'specialist' provision or equal opportunities monitoring, as 'loony', excessive and inappropriate.

The second factor substantially to affect the building of gay equality path-ways, particularly in authorities with greater normative commitment, highlights the interrelationship between environment and agency. I want to call this factor 'crisis/overload' to flag up its role in both naming, and guiding the response to, an environment in which organisational actors saw themselves as subject to an accelerating 'conveyor-belt' of priorities, across which impending catastro-phe loomed. Actors' perceptions – mythic or otherwise – that councils were facing immense crises, and that officers were being treated as 'in-trays' for the disposal of priorities to which politicians could not say 'no', meant that work seen by officers as unimportant, irrelevant, difficult or inappropriate was left to permanently 'pend'.

Crisis/overload in the late 1980s had an effect on the pathways constructed somewhat different from the logic of appropriateness identified above. While the former worked to stop some pathways from being mapped, overload/catastrophe created a gap between those pathways mandated and those actually built and used. While this frequently remained unjustified and the extent of non-action hidden, when pushed council officers argued that the scale of the demands on local government – and the crisis it faced – meant that gay equality could not rightly be accorded priority. In other cases the gap did not need to be rationalised through normative citations. To the extent that councils' gaze was fixed on the pressures and challenges of their wider institutional environment, they ceded the terrain of practical conduct, particularly in more marginal policy areas, to the logic of the doable.

The relationship of environmental conditions to the pathways of local actors was not solely a one-way process of the former structuring the latter. Although this did occur, the relationship was also more interactive as actors responded in different ways to the environmental conditions they confronted. Leaving to one side the personal workplace pathways of lesbian and gay council staff (see Humphrey 1999), I want to outline several other responses. To begin with, newly created organisational routes – in a constant state of flux and reformation – received uneven usage. Many council actors continued to treat them, as I have said, as inappropriate and illegitimate, requiring budgetary resources and time too scarce to provide. Instead, many policy-makers and front-line service deliverers embedded themselves within old, familiar routines. In some cases they also adopted hostile counter-pathways, such as pursuing excessive consultation with councillors, community representatives and officers known to oppose lesbian and gay work, in order to hinder policy implementation.

Specialist council staff, briefed to develop lesbian and gay work, did make attempts to connect their pathways with the routines of those who, while personally sympathetic, felt stymied by unconducive environmental conditions. Their efforts demonstrate the multiple ways in which routines can be seen as accomplishments (Feldman 2000), through the hard graft that goes into creating new practices, the ways in which they are solidified through use, and the attempt to achieve goals through the deployment of new routines. Specialist staff set up consultation meetings, rerouted responsibility and provided back-up for front-line work. Other strategies, such as building alliances with gay and lesbian community groups and leveraging senior political support, were also used to construct a more conducive interior environment, although they could do little to tackle directly the causes of feelings of crisis/overload and other exterior pressures.

Thanks to the difficulties they encountered, lesbian and gay specialist officers and committee members did not always, themselves, work through official pathways. In the London authority of Haringey, in the mid-1980s, the lesbian and gay unit bypassed the 'proper' managerial route of going through what seemed to them the 'black box' of the education department to write directly to headteachers about an equal opportunities curriculum fund that would support initiatives seeking to generate, among other things, positive images of lesbians and gay men (Cooper 1994: ch. 6). Someone leaked to members of the local Conservative Party the fact that the 'official' policy route, with its slow, highly cautious approach to policy development, had been circumvented. Drawing on the support of other opponents of municipal gay work, local Tories claimed that the council had abandoned its proper institutional pathways – procedural and substantive. The challenge rocked the council, which focused attention on the procedural irregularity, their degree of anxiety clearly influenced by outside perceptions of the impropriety of expanding pro-gay municipal pathways. Their actions, however, were not enough to cause the opposition to abate, an escalation which ended with the passage of s. 28 Local Government Act 1988,[4] a measure that sought to suppress the public, de jure creation of municipal pathways based on the incorporation of sexuality within liberal equality norms.

To the extent that s. 28 had a cooling effect on local government and its environment, particularly in education, was lesbian and gay municipal work any more effective in mapping and constructing new routines for residents beyond the town hall? Leaving to one side the impact of the policies on the voluntary sector and the new right, I want to suggest that new de jure pathways did emerge as a result of procedural changes. Local lesbian and gay activists, for example, found doors had opened, allowing them to participate in policy development and scrutiny through membership of council committees. At the same time, substantive policy changes facilitated the creation of new de facto routes as lesbian and gay residents (as well as visitors) became beneficiaries of revised anti-discriminatory policies, organisational funding and 'special' services. Yet while the pathways of some lesbians and gay men changed in response to new cultural initiatives, targeted employment opportunities and improved access to services, the extent to which a wider and more lasting impact on local pathways evolved remains uncertain. Council developments undeniably raised expectations among some lesbians and gays, facilitated the embedding of sexual orientation within prevailing equal opportunity discourses and helped to normalise gay sexuality as a publicly legitimate sexual orientation. In that sense, the council work of the late 1980s contributed to a shift in the institutional environment so that, by 2003, lesbian and gay anti-discriminatory work, particularly in relation to harassment and violence, had become an unremarkable staple of an equality/social diversity agenda (Cooper and Monro 2003). However, these changes were arguably due more to the publicised *mapping* of new pathways for officers, councils and communities than to their actual operation.

Some might argue that the decoupling of presentation from usage, so apparent in local government during the 1980s, was intended (see generally Brunsson and Olsen 1993: 39; Meyer and Rowan 1977). Many activists and commentators at the time pointed to the ambivalence of councillors and officers even in authorities pioneering lesbian and gay equality. From this perspective, the policies and mapped pathways were simply a symbolic offering to lesbian and gay activists and supporters, deliberately or complacently separate from the unofficial but utilised pathways which largely maintained the status quo. Leaving to one side the validity of this critique in a context where symbols and 'reality' cannot be meaningfully separated, could the council have created better utilised organisational pathways? To the extent that this required a more conducive municipal environment, I want to suggest that local government's capacity was limited. While internal environmental adjustments were possible, councils could do little, particularly in this area, to revise directly the wider environment within which they were embedded. To ignore the logic – the rationalities and pressures – generated by elections, media, and central government may have been temporarily possible, as certain more radical councils attempted to do. However, over any longer period, the membrane 'enclosing' local councils was too thin and permeable to enable a qualitatively different internal environment to survive.

While progressive councils in Britain were unable during the 1980s and early 1990s significantly to alter the institutional environment of local government, central government had more power in this respect, evidenced by its major statutory overhaul of the relationship between schools, local education authorities (LEAs) and service users during the same period (Bell 1999). These changes continued post-1997, under a Labour government seemingly committed also to testing, league tables and greater private-sector involvement in the management and support of schools (Beckett 2000; Hatcher 2000; Lushington 2000). I want to use here, however, for illustrative purposes, the earlier Education Reform Act (ERA) 1988, a signature Act that proved central to the wider strategy of transforming school environments and the education system more generally, mimetically according to the 'logic' and processes of the market (Ball 1990, 1993; Demaine 1988; cf. Hatcher 1994). By targeting both the educational environment and system, the Act helped to reconfigure the pathways of individual educators and schools despite considerable ideological and professional opposition to the reforms. Open enrolment, per capita funding and league tables, alongside the growing residualisation of local education authorities and the relocation of managerial control, substantially altered pathways pursued by schools and staff. Adapting to the new market environment, schools, for instance, adopted strategies to attract 'desirable' students (e.g., Gewirtz et al. 1993: 248), frequently in the face of their own ambivalence towards interschool competition and the unequal valuing of students (Ball 1993: 109).

Yet to suggest that the increasingly neo-liberal character of the educational environment during this period restructured pathways while leaving actors' subjectivity and desires untouched, underestimates the impact of the reforms. The process of being 'hailed' by market forces; the requirement of 'practical mastery' (Bourdieu 1992: 62) for success and even, in some cases, survival; the growing gap in function and perspective between senior school management and other teachers; and the new responsibilities placed on schools for economic and staffing decisions led many senior educators, over time, to integrate and *institutionalise* market imperatives within their reimagining of choices, obstacles and objectives (see generally Ball 1990: 15, 1993; Fergusson 1994; Hatcher 1994: 55; Hoggett 1996: 16; Marren and Levacic 1994). Yet, while retention of an unsullied, pre-market habitus proved close to impossible given the environmental conditions educators faced, actors' professional values and ethos did not undergo complete reconstruction (see also Clarke and Newman 1997: 99–100). Previously established norms and the educational field's general permeability meant that competing and contradictory norms could be held simultaneously (see also Cooper 1995b, c). Certainly, in the early 1990s, non-market discourses and practices, perceived by educators as lying at the core of their 'mission', such as equality of opportunity, continued to shape the de facto pathways that developed within schools.

What then needs teasing out is the way in which different institutional frameworks encompassing old and new norms (see Deem et al. 1994; Huckman and Hill 1994; Nixon et al. 1997: 21) became interwoven. This occurred both in the imagining of new pathways and in their creation by school-based actors. Gewirtz and her collaborators (1993: 241), discussing the impact of the 1988 reforms, quote the suggestion of one governing body chair, with a history of involvement in socialist education politics, that his school might usefully acquire a 'niche market', that would generate an inflow of students, by 'establish[ing] itself [as] doing well by black kids'. A second example drawn from my own research also demonstrates the improvised, and often unanticipated, character of the pathways generated as market norms intersected pre-existing equal opportunity discourses and practices.

It concerns the case of a London primary school head who, in 1993, turned down subsidised tickets for her pupils to attend a ballet production of Shakespeare's *Romeo and Juliet*. The reason reported to the saleswoman, in the course of an unsolicited telephone conversation, was that the school took equal opportunities very seriously, and *Romeo and Juliet* was a tale of exclusive heterosexual love (Cooper 1998a: ch. 5). The incident led to considerable controversy and enormous attention from the media, the local education authority and others (Epstein 1996). However, its particular relevance here is in the way in which it illustrates the complexity and unpredictability of paths taken in response to environmental changes. Jane Brown, the headteacher, was taken to task by her local education authority for suggesting that Britain's great, national playwright was heterosexist. However, the actual reason why she vouchsafed this claim may have had far more to do with her irritation at the unsolicited nature of the sales call than any deeper beliefs about Shakespeare. From this perspective, her reading of the play in terms of its equal opportunities breach constituted both an unplanned expression of annoyance and a means of putting the saleswoman off her stride in order to bring the conversation to a close. In the 1980s and 90s devolving budgetary powers, in a general context of increased marketisation, placed new, intensified pressures on school managers to deal with salespeople. The emergence of idiosyncratic as well as the more usual pathways, and the norms on which these drew, reflected the tensions and frustrations of educators, who, among other things, were driven to find ways of discouraging and terminating unwelcome solicitations.

Conclusion

My starting point for this chapter has been a challenge which has long confronted many political movements: what strategies and political changes offer the best means of creating new, sustainable routines? Government, community or pressure group exhortations to be vegetarian, recycle refuse, take public transport (rather than drive), work co-operatively and non-competitively, boycott supermarkets, and challenge racial, gender and sexual stereotypes may prove successful for brief moments. However, even if ideological commitment is sustained, new practices may well not endure if the cultural, economic and social costs are too great. This can prove true even when public authorities are the architects. As the example of lesbian and gay local government initiatives demonstrates, authorised pathways may fail to be built or may become quickly overgrown as de facto routes take other directions. Even orders and instructions to build and inhabit new paths may fail to achieve their objectives to the extent that subordinate actors are able and choose to 'environmentalise' the demands placed upon them.

This failure of new temporal trajectories to become embedded matters for reasons I explored at the start of this chapter and which I pick up again in the chapter that follows. Progressive politics frequently focuses on achieving reform and change. At government level, the symbolism of reform may be all that is really sought. However, to the extent that more egalitarian relations are desired, sutured to the circulation and citation of different normative principles, pathways that are embedded and that, as a result, do not fade quickly may be called for. In the second part of this chapter, I explored two attempts to create such counter-normative pathways, first, where major environmental changes were not introduced, and second, where they were. However, as my example of central government's educational reforms reveals, even where major environmental changes are introduced, the de facto pathways to emerge cannot be fully anticipated or easily contained. In part, this results from actors continuing to operate, at least for a while, in accordance with 'strings of moves' based on previously internalised norms, values and beliefs. More generally, it reflects the heterogeneity, or messiness, of many social environments in which divergent norms and other pressures pull conduct in competing or antagonistic directions.

The intellectual terrain of messiness and incoherence has been well traversed by postmodern scholarship, and I do not want to retread it here. What I do want to highlight, however, is the way in which a messy environment can facilitate and even sustain a range of different pathways. The lack of a snug or ergonomic fit between pathway and landscape may be for many reasons. Meyer, Scott and Deal (1983: 664–5) refer to 'buffering' – a process where decoupling, concealment and insulation are used to keep paths away from clashing practices and environments. While the focus of Meyer and his colleagues is on organisational processes, the concept of buffering can also be used to identify the way in which preferences and norms are structured so that people do not seek or desire pathways unavailable to them. When these processes fail to work effectively, and when gate-keeping also breaks down, pathways open up, yet users remain subject to the gravitational pull of other, competing 'logics'. This is the 'dilemmatic' environment, illustrated, for example, in the situation facing many 'working mothers' within modern, Western societies. Here, whichever trajectory a woman chooses – remaining in full-time employment, moving to part-time paid work or bringing up children full-time – she confronts the economic, social and cultural pull of her alternatives.

Bonnie Honig (1996: 259) develops a similar argument in her discussion of 'dilemmatic spaces'. Honig uses the term to emphasise the 'ineradicability of difference from identity'. In other words, conflicts and difference do not simply exist outside the subject, they also penetrate it.

[T]he circumstances of their subject constitution position all moral subjects in... *dilemmatic spaces* ... [H]er agency itself is constituted, even enabled – and not simply paralyzed – by daily dilemmatic choices and negotiations ... We ought not to think only in terms of dilemmas as discrete events onto which unitary agents ... stumble ... but perhaps also in terms of ... spaces that both constitute us and form the terrain of our existence.

(Honig 1996: 259)

Although the convergence of prevailing norms, social relations and institutional structures may cause certain routes to dominate, the heterogeneity of our social landscape, and the contradictions and dilemmas experienced at many social locations, cause criss-crossing pathways to proliferate. These include routes that are widely or more narrowly carved, with considerable or meagre impact, and more or less longevity, which conform to or contest prevailing environmental rationalities, and which are well-defined or as yet unfinished. At the same time, we are currently witnessing, in Britain and elsewhere, pressures to make social and institutional environments more coherent. At a theoretical level, liberal scholars such as Michael Walzer have argued for a model of relatively discrete fields operating according to their own specific norms. Politically, the move against messiness is also apparent. One site is the intensified mobilisation of normative principles of proper place and the deployment of nuisance claims against homeless people, travellers and other subordinate constituencies who 'hang around' (see chapter 6), and who thereby induce concern that the social environment will be 'messed up' for others. Another is the emphasis on networking between institutions, and political concepts such as 'joined-up thinking' which seek to establish environments that produce or make possible a 'coherent' series of authorised pathways.

Yet, as my emphasis on the value of counter-normative routines suggests, I do not want to celebrate, for its own sake, environmental messiness, incoherence and the proliferation of pathways. The organisation of paid and unpaid domestic work, and its consequences for women (and men) involved in bringing up children, is, from a progressive perspective, far from ideal. The fact that it may function for some as a 'messy' or a dilemmatic environment in which different, 'conflicted' choices can be made does not make it inherently preferable. In chapter 4 I argued for an equality politics based on undoing relations of gender, class, sexuality and race (among others). Arguably, the environmental changes this process would generate would eliminate certain de facto and de jure pathways (while also creating others). Again, the possible decline in messiness this might bring is not itself a reason to reject such a political project.

At the same time, in a political context where little institutional commitment exists to the radical undoing of relations of inequality (beyond the most formal manifestations of such), environmental messiness may have something to offer. Bonnie Honig (1996: 260) argues that the ungovernable, undecidable quality of dilemmatic spaces enables them to function as spaces of political engagement and empowerment. The dilemmatic environment not only maintains pathway diversity, but, because no route is without apparent costs, a messy environment generates and sustains uncertainty, reflection and choice. In particular, the anti-hegemonic, with its necessary critical, creative and spontaneous edge, survives in those spaces of environmental contradiction. But dilemmatic spaces can also sustain counter-normative choices which eke out a meagre survival as marginal, but just about viable, practices.

In the chapter that follows, I explore the difficulties that confront attempts to create enduring counter-normative pathways further by exploring the situation of prefigurative communities and networks – collectivities that seek to live out, as a current reality, aspirations for a different society. I suggested above that creating a conducive environment within current liberal societies for the enactment of more radical pathways can be difficult to bring about, given the opposition of dominant forces and institutions. However, one site where environmental change is possible is in smaller communities and networks which seek to create a controlled space of collective self-governance. But how possible is it to create an impenetrable space which allows different environmental conditions to prevail and thrive? Are such spaces also messy and heterogenous, and what implications does this have? Finally, to what extent do such communities impact, in progressive ways, beyond their own boundaries? These questions circulate through the penultimate chapter of this book.

Notes

1. At the same time, if people walk across fields or through a wood following their own personal inclination, if they ignore the logic or biases of the terrain (or if none exists), no path may be established. This right or freedom to roam evokes a libertarian postmodernism in which the social neither shapes nor is brought into being by common individual conduct. Thanks to John Clarke for suggesting this point.
2. This process is also complicated by the fact that laws often work more effectively when they change the actors' environment, for instance, through their normative impact or reallocation of resources, instead of confronting actors with new prohibitions or directives that appear simply to reduce discretion (see also Godard 2002).
3. For details of the primary data on which this discussion is based, relating to lesbian and gay equality policies in local government, see Cooper (1994).
4. Repealed in 2003.

8

Safeguarding community
pathways: 'possibly the happiest
school in the world'[1] and other
porous places

Processes of convergence and accommodation, whereby counter-normative
practices lose their sting as they become embraced by the status quo, have
been praised by those who see in this the productive workings of progress,
consensus and reform. More radical critics, meanwhile, have condemned these
shifts as assimilatory and co-optive. But what alternatives are there? How possible is it for oppositional practices to endure – given the relentless tug of
the mainstream? In the previous chapter I addressed the difficulty of maintaining oppositional practices in an inhospitable environment. The challenges
this difficulty generates have been tackled in several ways, a major one being through attempts to change the wider environment, as I explored in relation to British government education policy of the 1980s (see chapter 7).
My focus in this chapter is different. While some have derided it for being irrelevant and self-indulgent, creating bounded communities with their
own interior environments provides one way of protecting counter-normative
pathways.

My reason for focusing on counter-normative communities and boundary
techniques is threefold. First, I am interested in exploring the *materialisation*
of counter-normative principles. I have argued that normative principles do not
exist simply as claims or arguments but are embedded within, and can be read
off from, social practice. Counter-normative communities provide an important
site for examining attempts to live out alternative interpretations and articulations of normative principles such as equality, fairness and entitlement. Unlike
other political movements, such communities seek to generate new pathways
not simply through mapping – the discursive envisioning of alternatives – or
by campaigning for their construction, but also through their usage (see also
Lacey 1998, ch. 8). In the previous chapter I considered the construction of organisational pathways to 'deliver' change to others; in this chapter my concern
is largely with pathways that actually embody new norms.

Second, counter-normative communities offer a site through which to explore the *reproduction* and *maintenance* of alternative norms and social relations. Social movement activity and much political theory take, as their focus, the introduction of new norms, institutional processes and practices. It often seems to be assumed that once these changes have been introduced they will simply and unproblematically endure. But, particularly when changes are introduced diachronically, that is at different times, rather than through a single, integrated revolutionary rupture, most alterations and revisions are likely to confront an incompatible or, at best, awkwardly compatible environment (see also Lowndes 1997). Focusing on alternative communities makes it possible to examine social microcosms where radical changes have been introduced synchronically. This might be through the symbolic rupture, epitomised by 'free states' and temporary autonomous zones (TAZ) (e.g., Bey 1991; Seel 1997). Hakim Bey (1991: 101) describes the TAZ as being 'like an uprising which does not engage directly with the State, a guerrilla operation which liberates an area (of land, of time, of imagination) and then dissolves itself to re-form elsewhere/elsewhen'. Communities created to resist particular policies, such as new road building, may have no intention or desire for permanency, but to the extent that they temporarily reproduce daily life, and do so in ways that differ from the mainstream, communities such as these illuminate the achievement and maintainence of oppositional practices and norms.

Third, progressive communities offer a site through which to explore the interface between equality and other social norms. Earlier I argued that diversity and equality should not be seen as inherently antagonistic, although they are often presented and articulated as such. This is particularly apparent in work that draws an equivalence between diversity and negative freedom and juxtaposes it against equality as sameness. Considering communities oriented around notions of equality thicker than simply procedural fairness allows us to explore the kinds of diversity produced. However, in considering the relationship between equality and diversity, I want to flag up a point, spelled out in more detail later on, that the communities discussed are not counter-normative in every respect. Close study also shows the ways in which they converge with dominant norms, relations and institutions. One question underpinning this chapter is whether such convergence is necessarily bad. While it may limit the extent to which communities showcase alternative lifestyles, it may have other advantages, as I go on to explore.

Sustaining communities through their boundaries

[A]ny attempt to inspire people with new visions, with new definitions of what is possible for themselves as individuals and for society in general, any attempt to inspire people with new criteria by which to adjudge the quality of their lives and relations with others, is a fundamental part of any revolutionary process.   (Rigby 1974: 147)

As I discussed in chapter 7, counter-normative pathways encounter a range of difficulties in attempting to be sustainable. Prevailing social practices are able to survive largely because their conditions of existence are reproduced. For example, the consumer demand necessary for commercial profitability is maintained by property laws and credit systems, as well as by marketing practices, welfare policies and wage-based employment relations. In contrast, the conditions of survival for oppositional pathways in an alien environment seem far less assured. Even when they are developed within organisational settings, such as the lesbian and gay equality policies pursued within British local government (see chapter 7), the interior environment may be far too saturated or dependent on a hostile wider environment for counter-normative pathways to endure. Creating bounded communities, where different environmental conditions can effectively operate, offers one way of providing a space within which a form of counter-normative social reproduction can occur – where resource allocations, institutional frameworks, discourses, norms and conduct bolster and sustain each other, and where progressive pathways are constituted as rational and benign rather than absurd, tiresome or disruptive. This does not deny the extent to which oppositional practices may be motivated by, and produced out of, hostility to mainstream or dominant processes. The centrifugal or 'exiting' effects of dominant social relations and practices emerge very clearly in discussions of communes and other intentional communities (Kanter 1972: 7). However, while opposition may provide the impetus for quitting the mainstream, without any form of 'protection' the sustainability of particular alternative practices, over time, tends to be weak (see, e.g., Pepper 1991).

The process of constructing barriers against the 'real world' is a familar technique with a long, well-established history, deployed by traditional and religious, as well as radical and socialist, communities (see Kanter 1972; Metcalf 1995; Pepper 1991; Sargisson 2000). Indeed, boundary building is, arguably, an inherent aspect of all communities and societies, a social dynamic that produces, and works through, the division between inside and out, between that which is valorised and empowered, and that which is devalued, marginalised and disempowered. The creation and maintenance of physical and legal boundaries, from national borders to the emergence of gated communities and the privatisation of common spaces (see Judd 1995), has proved to be a central aspect in the reproduction of inequalities. However, communities do not have to operate in this way, and with these effects. In this chapter I focus on prefigurative and progressive communities and the range of boundaries on which they draw. Boundaries can be normative, physical, social, economic and legal. And it is not only physical boundaries that can take different forms. Metaphorically, others too operate as walls – clearly defined, visible, fixed and relatively impermeable – as membranes – far more flexible, transparent and porous – as gates – where passage is confined, regulated and scrutinised – and as canal locks – devices that mediate inside and out.

The character and form taken by boundaries depend on both the reasons for their establishment and the external challenges encountered by a community over its lifetime. Discussing the Anabaptist Hutterite communities of North America, Barry Shenker (1986: 116) argues that 'economic strength' enabled not only economic, but also 'social and political independence, a freedom from external coercion, the absence of which would not only mean undue ideological influence from outside but also a loss of faith on the part of members in their ability to control their own lives'. Boundary techniques may also change over time. For instance, nation-state paraphernalia (passports, flag or constitution) may be deployed by communities initially as symbols of their opposition to government action and of their figurative intention to secede (Seel 1997: 109–10). However, subsequently, they may play an important role in contributing to the actual material of the boundary in an attempt to maintain territorial and cultural separateness or autonomy. Similar adaptability can be seen in the role of normative beliefs. Although oppositional norms may be mobilised and flagged up to generate the momentum for change, they may not be explicitly needed, in the early stages, as a way of walling off the mainstream. However, over time, as the material, emotional, ideological or pragmatic aspects of the mainstream exert greater gravitational pull, community norms and commitment may become relied on, and mobilised far more explicitly, to ensure the community's survival (see Kanter 1972: 65).

In general, the role of boundaries is threefold: to keep the status quo at a geographical, socioeconomic or cultural distance, to intensify members' commitment and familiarity towards localised practices, and to negate the attractiveness or desirability of the status quo (see also Rigby 1974: 140). In previous chapters I have referred to different boundary processes from the revision to inside and out engineered by the legal introduction of lesbian and gay marriage to the eruv. Crafted out of the existing landscape and completed by the installation of figurative doorways, the eruv functioned to delineate a community's limits, while simultaneously facilitating 'normal' conduct – carrying and pushing on the Sabbath (see chapter 2). In the discussion that follows, I consider three quite different instances of boundary construction. The first concerns the geographical and gender-based separatism of the Greenham Common Women's Peace Camp, a temporary, action-oriented community that existed in Britain during the 1980s.[2] My second example addresses the counter-normative commitment and reformulated self-interest evidenced in local exchange trading systems (LETS). Established in many countries during the 1980s and 1990s, LETS offered a formalisation of extended neighbouring based on new, local currencies. My third example is that of progressive or 'free' schools, where I consider the case of Summerhill School in eastern England, one of the oldest and best-known alternative schools worldwide, famous, in particular, for its commitment to democratic self-governance and children's freedom.

These three studies provide points of contrast as well as similarity. Differences between them can be seen in their aims: to protest, to enable trading and to educate; in their relationship to law: where Greenham's defiant disregard contrasted with LETS location within a space of legal possibility and with Summerhill's official legal status; and in their physicality: Summerhill and Greenham benefited from a clearly defined territory, while LETS have lacked their own communal space. In terms of similarities, all three communities demonstrated prefigurative qualities in the sense of modelling alternative, more egalitarian forms of social relationship (although the extent of their orientation to wider change varied). At the same time, their 'realism' distinguished them from other, more Utopian communities. None of the three, for instance, sought to create 'new and better people'. Their aim, rather, was to develop relations and spaces in which members could come to know and express their 'true' interests and needs (e.g., Neill 1968: 111). Additionally, neither LETS, Greenham nor Summerhill sought to construct impermeable boundaries; all were dependent on a constant movement – particularly an influx – of people. They were also not 'movements of withdrawal' (Kanter 1972: 214). In the case of Greenham, the aim was to change wider military policy through temporarily setting up a peace camp. LETS' *raison d'être* did not revolve around political protest, but they were also not intended as a form of exodus from mainstream life, since members participated in LETS alongside other social and economic relations, as I discuss further below. Finally, Summerhill, like other progressive schools, aimed to create free, self-governing young adults who, on leaving, would function effectively in the wider world.

The Greenham Common peace camp

Sustaining oppositional culture . . . depends on a multiform practice of community building and identity formation, that creates affective bonds between members and nurtures the connectivity of community life while also contesting hegemonic constructions of subaltern subjectivity.   (Carroll and Ratner 2001: 608)

The Greenham Common Women's Peace Camp was one of the most prominent and influential British feminist communities of the late twentieth century (Blackwood 1984; Cresswell 1996; Roseneil 1995, 2000). Established in 1981, the camp was the culmination of a peace march led by women that began in Cardiff and ended at the US military base just outside Newbury in southern England. Greenham began life as a mixed camp. However, in February 1982, women members decided, for symbolic, practical and strategic reasons, to become a women-only space (Jones 1983; Liddington 1989: 235–6). By placing women at the centre of the struggle against nuclear weapons, Greenham, like Seneca women's peace camp founded in 1983 in upstate New York, 'foreground[ed] gender as a major issue in the critique of the militaristic–patriarchal–nuclear establishment' (Linton 1989; Krasniewicz 1992: 45). In this way, the peace camp was not just about nuclear weapons or the presence of the US military in Britain. In the course of its lifetime, Greenham also functioned as a prefigurative feminist community committed to collective forms of subsistence, equality, non-hierarchical norms and participative democratic values. Sasha Roseneil (2000: 124), who both studied and lived at the camp, wrote, 'There was a strong discourse of equality and anti-hierarchy, and an anarchist rejection of leaders and "stars". . . . Greenham's democracy of the present demanded that the opinions of the newly arrived be treated with as much regard as those of the long-term residents.'

However, the description of Greenham (and women's peace camps more generally) as prefigurative is not uncontroversial. Feminist critics at the time argued that the values and practices adopted at Greenham privileged the agendas of white, middle-class women and perpetuated traditional and stereotypical depictions of femininity (Amos and Parmar 1984; Onlywomen Press 1983). In particular, critics pointed to the discursive deployment of motherhood, and camp members' emphasis on women's 'special' relationship to peace and to the future (see Liddington 1989). This was especially noticeable in several mass protests, when women pinned symbols of mothering to the wire fence – including photographs and clothing of their children – mobilised 'feminine' forms of mysticism, and used song, chants and playful actions to rebuff the hostility of military, police, bailiffs and local residents.

At the same time, despite the feminist criticism Greenham faced, the camp provided an important space for the development of protest-based pathways. These pathways articulated, and were embedded within, countervailing conceptions of proper place that challenged the base's legitimate and natural anchoring within the landscape (see Cresswell 1996; also chapters 5 and 6). This counter-normative framing extended into, and was accompanied by, a refusal to respect the propriety of legal rules and procedures. As women were prosecuted for criminal trespass, breach of the peace and criminal damage, camp members and supporters used song, wailing, political speeches and mass attendance to carve out new de facto routes through the legal system and juridical spaces in which they found themselves. As one participant described, 'Some women . . . abandon the legal language and break into song. Others shout out that magistrates must listen . . . Our court cases often end in pandemonium'

(Johnson 1986: 176). The women charged did not simply reverse the prosecution's claims of out of placeness, although they did do just that, arguing that it was nuclear weapons that were 'breaching the peace' (see Johnson 1986). More generally, in the face of the court's individualising tendencies and pursuit of solemn authority, defendants and their supporters brought pathways of passion and solidarity into the courtroom.

Greenham's counter-normative pathways were not restricted to opposing war and formal institutional processes. As many commentators and participants have discussed, Greenham also provided an environment within which alternative gender principles and organisational norms could be established (e.g., Roseneil 1995, 2000). Some counter-normative pathways emerged out of the conditions in which the camp existed. Living in tents made of branches and plastic tarpaulins, in the cold, damp and mud, without plumbing, electricity or gas, disciplinary norms that conventionally regulated women's appearance and body smells became impossible to sustain (although a counter-cultural Greenham aesthetic did also emerge) (Kirk 1989; Roseneil 2000: 95). Women were also obliged to utilise and develop non-traditional manual skills to sustain the camp without the participation of men. Other normative principles, such as the refusal to institute a leadership structure and an open approach to domestic chores, emerged as an intrinsic part of the camp's (anarcho-)feminist ethic (Roseneil 2000; cf. Linton 1989 on Seneca peace camp). At Greenham, household responsibilities were to be carried out voluntarily and willingly or not at all. As a consequence, work to reproduce social life became very unevenly taken up (see Liddington 1989: 282–3; Roseneil 2000: 100–4).

The application of negative freedom to domestic life at the camp provides one instance of the way in which diverse pathways emerged out of a commitment to individual equality. Social diversity was also acknowledged in the camp's rejection of heterosexual hegemony. Krasniewicz (1992: 67) makes a similar point in relation to Seneca: 'Many women at the encampment considered lesbianism an important aspect of their alternative, non-patriarchal world. Since the encampment was designed to be a physically, socially, and spiritually safe place, women were encouraged to express all aspects of their femaleness freely. For many women, identifying as a lesbian was the most important aspect of this revised identity.' Although some critics claimed that lesbianism had become a Greenham norm that deauthorised heterosexual desires, others saw Greenham as offering a space that dismantled heterosexuality's privileged status without putting an alternative sexuality in its place. One way in which this was sustained was through the creation of sub-camps at each of the base's gates (Roseneil 2000: 76–84). This enabled different norms to flourish, and to some extent to be buffered from each other at least in daily life, through micro-communities differentiated by space and ethos.

In many respects Greenham peace camp provides a good example of a counter-normative community. Yet while the status quo provided the camp with its impetus to become established, it also exerted a gravitational pull on the camp's ability to endure. This gravitational pull took several forms. It included the lure of better living conditions – of comfort, food, hygiene and warmth. While many camp members had limited financial means, few would have found their living conditions at 'home' as basic as those experienced at Greenham. Traditionally, the desire to secure 'better' living conditions has proved to be a major factor in causing people to leave communes where the limited wealth produced by ethical work or self-sustainability is equally shared. It has also led some communes, over time, to prioritise profitability, material comfort and desert over other motivating values (see, e.g., Pepper 1991: 192–8). While material conditions at Greenham did not bear the same significance as in more permanent 'quality of life' communities, privations intersected other centripedal tendencies exerted by the mainstream, such as familial and emotional ties, living 'within the law', and the normative construction of 'sensible', 'desirable' living. These complemented the centrifugal tendencies of the camp in which boredom, fatigue, demoralisation, the strain of collective living, and relentless pressures from harassment and abuse worked also to propel camp members away.

Against these pressures Greenham utilised, not always with conscious intent, mechanisms to safeguard camp practices and structures. I want briefly to outline four: space, norms, pleasure and transience. From the outset, physical separation – the occupation of land – proved crucial to creating a space within which different everyday pathways could be adopted. Living at a geographical distance from the mainstream, in a space defined by other imperatives, made it possible to pursue confident, visible lesbian relationships, non-traditional skills, physical assertiveness and the marginalisation of feminine aesthetic and domestic norms. However, while separate physical space proved to be a crucial factor in enabling alternative community practices, Greenham proved a highly permeable space. This left it far less successful in protecting the camp from the environmental frictions caused by hostile others, particularly those others with legal rights of entry, such as bailiffs and police officers, or with sufficient physical proximity (from their own spaces), such as soldiers on the base and nearby residents, to carry out verbal or visual acts of obscenity or malevolence.

Political commitment and normative beliefs were also central in maintaining Greenham as a space within which different practices could prevail and in undermining the mainstream's gravitational pull. Not only was this pull, with its logic of rational action and the good life, defused since its terms largely did not appeal to camp members, but mainstream values were also discursively cited as negative principles to strengthen the camp's motivation to continue. At the same time political rationalisation and commitment provided, at best, a partial barrier against less normative centripedal and centrifugal pressures. Many protest camps have found that they cannot survive, over time, on committed beliefs alone. One form of supplementation to augment commitment is the deployment of fun, sensuality and creativity. The playfulness and theatricality of Greenham actions, such as the costumed teddy-bears' picnic on the base,[3] proved to be an important way of sustaining commitment in the face of internal privations, external enticements and mainstream rationalities. According to Roseneil (2000: 133–4), enjoyment was a guiding principle at Greenham: an assertion that women were not sacrificing themselves for the benefit of others. In her discussion of the mystical and spiritual aspects of camp actions which played a similar role, Liddington (1989: 236) writes that 'extraordinary times call for extraordinary response; and ritual, symbols and incantations soon assumed a vital role in sustaining such an unlikely being as a women's peace camp outside a nuclear base'.

Given the centrality of political commitment, it is perhaps unsurprising that a major reason for the camp's eventual decline was the complex confluence of new political currents, in the late 1980s, both within and beyond the women's peace movement. However, in a sense, the camp was intended as a temporary measure from the start – an identification that played an important role in keeping the gravitational force of the mainstream at bay. Since the camp was not meant to provide a permanent home, domestic (and other) expectations were lower, and the constant evictions, although exhausting and discouraging, were part of a struggle – crisis points that demanded a reaction (or resilience) – rather than demonstrating the impossibility of sustaining a desirable permanent settlement. The temporary quality of the camp also made a turnover of women possible and legitimate. Liddington (1989: 276), in her discussion of Greenham, suggests that there were cycles in the lives of the women there, where actions and evictions signalled transitional moments in which some women would leave and others arrive. The possibilities of renewal that emerged from the juncture of political commitment and geographical permeability allowed Greenham to counter potential downward spirals of frustration, demoralisation and fatigue.

Local exchange trading systems

My second case-study concerns local exchange trading systems. LETS began on Vancouver Island, in Canada, in the early 1980s, as a way of recharging an economy hit by heavy unemployment thanks to the closure of the local mine. LETS subsequently spread to the United Kingdom, Australia, the United States, Japan, New Zealand and Ireland, among others (Croall 1997: 14–15). LETS function as small-scale webs of intersecting pathways, in which de facto 'strings of practices' riff around the basic de jure path. This basic path plots the relationship between exchange practices and an accounting system, enabling goods and services to be traded in a local (non-state-based) currency. LETS generated considerable interest through the 1990s (see Barry and Proops 2000; Bowring 1998; Lee 1996; North 1999; O'Doherty et al. 1999; Purdue et al. 1997) as an imaginative means of reconnecting economic relations to local social networks in ways that strengthened both. LETS also illustrate the complementarity of diversity and equality. The starting point for LETS is the equal status and trading position of all members. Yet this does not mean a celebration of sameness. Aside from the diversity of beliefs and values among members that I return to below, LETS' success depends on the range and breadth of skills, offers and wants that members possess. In this chapter I want to use LETS to explore the difficulties faced by attempts to sustain counter-normative practices. But first I want to address the claim that LETS are not really counter-normative at all, given the way in which they dovetail with dominant socioeconomic relations.

Arguments about LETS' compatibility or convergence with the mainstream have been made in several ways. Commentators such as Williams (1996: 89) have argued that LETS constitute a rational response to the problems capitalism creates for those it economically disenfranchises. In the United Kingdom LETS have been supported by some local authorities as a means of community and economic development in deprived areas. In Australia government ministers saw LETS as a way of helping the unemployed to develop and practice the skills and values (including self-discipline) necessary for re-employment (Bowring 1998: 100; Croall 1997). More generally, LETS can be read as part of a broader project of social inclusion intent on integrating marginalised citizens within mainstream social relations.

LETS may also be interpreted as compatible with, rather than as challenging, dominant social relations and dynamics in several other respects. Kanter's (1972: 153) use of the term 'isomorphic' is useful in this regard in signalling the extent to which LETS can be mapped on to the wider environment through a shared language, set of symbols, media of exchange and roles. For at heart, LETS are about trading and markets. Although they do not reproduce capitalist relations of accumulation and exploitation, they incorporate many of the economic instruments found within capitalism, such as labour, exchange, consumption, commodification, currency[4] and accounting. As a result, although LETS encourage members to construct offers around things they enjoy doing, the logic of the market means people are propelled into providing the kinds of services or goods for which there is a demand.

LETS also reflect the individualism of trading, in which exchange relations take place between two parties (although in some schemes group work projects may occur). As such, they offer a more formalised version of neighbouring in which currency is used to organise, facilitate and extend the range of 'favours' otherwise carried out between people living in proximity to each other. LETS reflect a habitus which prefers reciprocality to one-way favours, and knowing 'where one stands' over informal obligations. Within this context, LETS extend the intimate/impersonal dynamic (see chapter 3), as commodification incorporates previously unpriced favours: LETS economies typically revolve around services and goods (such as home-made produce, discarded furniture, lifts, garden help, small amounts of typing) outside the formal economy (although LETS do include other transactions). The power of the intimate/impersonal is also evident in the kinds of social interactions which remain outside LETS relationships. Although LETS are used to structure exchanges with adjacent 'strangers', they have largely not been used to organise intra-household transactions, while sexual exchanges are formally or informally ruled out.

Finally, LETS reveal the capacity of dominant socioeconomic inequalities to permeate local systems. In the United Kingdom a handful of LETS were specifically designed to undo inequalities seen to derive from a lack of confidence and skills. Croall (1997: 57), for instance, describes a women's scheme in Glasgow which aimed at encouraging women to learn non-traditional skills. A scheme also operated in Leicester for Asian women, and several schemes were developed for users of mental health services. However, in the main, open LETS membership and minimal scheme governance generated a 'hands off' response towards organising principles of inequality. The effects of this were evident not only in who did not join but also in the distribution of offers, wants and value. According to Aldridge and her colleagues (2001: 576), 'the negotiation of value . . . is shaped both by LETS members' positions within the formal economy and their relations to each other. As such, values are often transported from the formal to the informal economy.'

In flagging up the less transformative or radical aspects of LETS, my aim is not to suggest that they are in fact reactionary or conservative institutions, nor that the benefits and limitations of LETS should be weighed up and balanced. Rather, as I consider further below, LETS' openness and partial convergence with the mainstream is important when considering its wider impact. Yet, the isomorphism and permeability of LETS needs to be addressed in the context of its capacity to offer a counter-normative market structure. On a small scale, the institutional LETS framework enabled new, more equal pathways to emerge that would be hard, if not impossible, without it. Multilateral exchanges depend on technologies of accounting or currency, and although these technologies may be normatively thin in themselves, in the LETS context they facilitated pathways oriented around collaboration, commitment, privacy and equality.

LETS' counter-project takes three major forms. First, LETS exclude exploitative wage–labour relations, as people gain LETS credits from the work they themselves do (although the involvement of some local companies in LETS complicates this assertion). LETS also avoid the intensification of inequalities that comes from the accumulation and reinvestment of profit. While members can build up credit balances, there is little they can do with earned currency other than spend it on LETS goods and services. Second, LETS operate with pricing mechanisms that trouble, to some degree at least, conventional notions of value (Lee 1996: 1382–3; cf. Thorne 1996). While a substantial number of LETS have allowed members to trade at the going price for their goods and services, other LETS adopted different approaches to the contentious question of value (see Aldridge et al. 2001). In some schemes a uniform rate was adopted, encouraged or deployed as a benchmark, usually based on an hourly rate (see also Seyfang 2001: 589). In other schemes, caps and floors controlled the range of pricing differentials.

Third, LETS have sought to transform the character of economic transactions, and to re-embed more symmetrical and egalitarian exchanges within local community relations.[5] LETS exchanges are different from conventional market ones in that trading relationships are intended to be multi-shot, with a stress on the performance of social bonds. Members whom I interviewed highlighted the way in which the 'purchase' and delivery of services, especially manual and domestic ones, were carried out according to an ethos of mannerly reciprocity. As one woman I spoke to said, 'I'm more flexible and open to the person whose needs are being met [when providing a LETS service]. I never know when I'll want them to do something . . . I don't rush people because I know they have a real job . . . I'm more respectful of people's time than when I pay.'

Embedded within a broader history of alternative economic experimentation, LETS offer a trading structure and network of practices that challenge the assumption that economic relations must necessarily involve alienation and exploitation. LETS' viability 'pinches' the confidence of capitalism which relies on communities' dependence on wage–labour relations. The abundance of LETS currency, the multilateral quality of exchanges and the emphasis on developing social relations through economic ones provide a workshop in which normative qualities of trading based on generosity, responsibility and equality can be tested and developed. But if LETS provide an alternative, more moral framework for economic relations, to what extent are they vulnerable to the norms, rationalities and configurations of the wider environment? Do participants encounter the gravitational pull of the mainstream in ways that counter the pathways LETS seek to create? To what extent have LETS constructed or deployed boundary techniques in response?

The gravitational pull of the status quo, in relation to LETS, has operated in several ways. At the level of inducement or incentive, the main effect has been on those members (or potential members) whose skills or produce attract a higher price within the mainstream economy. For LETS to sustain economic structures that raise the price of undervalued manual work, other work has to be priced below its mainstream market equivalent. Members of some skilled, highly sought-after trades, particularly plumbers, builders and car mechanics, have been underrepresented in many LETS,[6] despite the high demand for such services from LETS members, because, it is generally felt, they can earn more within the mainstream economy. Alternatively, these and other services may be offered by members in limited quantity, since their priority is to sell their skills – or fulfil employment contracts – within the better paying dominant economy.

The centripedal pull of mainstream markets also affects the temporal trajectories of LETS members. Croall (1997: 19) describes a career progression in which people start by offering a skill in LETS, create a market for them and subsequently shift to offering the trade half in LETS currency and half in sterling. Eventually they may become sufficiently established to stop offering their skill in local currency altogether. A related trend, witnessed from my research, was a shift in the services members offered from those that resembled their mainstream job to other, lower valued tasks. Thus people who initially offered different kinds of therapy changed to offering cleaning, transportation or weeding to avoid the substitutional effects caused by people paying reduced rates in a local currency rather than the going rate in sterling.

While the mainstream economy holds out inducements that may limit the participation of higher valued service-providers, it also creates other pressures. These counter the capacity of LETS pathways to become embedded and widely used. One is that of space. Without a single spatial domain, LETS exchanges tend to occur in the households of members. This makes LETS vulnerable to geographies of drawn-out travel time, and has meant that people living on the physical peripheries of schemes are less likely to participate. Geography has also had an effect, whose scale is hard to determine, on attempts to include people who live, from the perspective of other members, in unsafe spaces. During my interviews several members referred, equivocally, to entering or leaving their car in districts perceived as more than usually dangerous. A second factor is time. LETS are largely circumscribed by the residual space left open for hobbies in the lives of many full-time employed people. Although LETS draw people with more 'free' or uncommitted time, in order to expand they also need to be able to attract people currently in full-time employment. However, the limited market-range of most LETS makes it (almost) impossible to exit completely from mainstream economic transactions (see Lee 1996: 1385), even if participants are willing to forfeit a higher standard of living.

LETS also confront wider norms that govern consumer relations – oriented in the West around efficiency, transactional ease (such as booking and price), quality and minimal purchaser obligations (in terms of conversation, friendliness, reciprocation). To the extent that these norms prevail as expectations and standards, they reduce the desirability of LETS, since the latter are seen as less reliable,[7] more awkward to initiate, and more demanding on the purchaser. But LETS also rub uneasily up against the prevailing ways in which economic relations are understood. Although mainstream transactions are frequently seen as competitive and antagonistic, they are familiar. Consumer magazines and television shows advise buyers and sellers to acquire mastery in the complex exchange pathways they inhabit to reduce the likelihood of losing out, being cheated or appearing foolish. In contrast, LETS ask people to believe that economic relations can be different. However, to the extent that people bring with them a habitus structured by antagonistic trading relations, the unfamiliarity, minimal regulation and lack of sanctions within LETS can generate heightened levels of transactional anxiety.

The impact of the pressures exerted on LETS by the wider environment and by the habitus people bring to the scheme raises questions about community boundaries. What mechanisms have LETS drawn upon to protect their environment and pathways against the gravitational pull of the mainstream, especially in a context where most, if not all, LETS members are boundary actors, straddling and participating in different economies? And do these mechanisms encourage new people to join? For it is arguable that the mainstream's main impact has been less on the exiting of existing LETS members (although this has occurred quite extensively) than in the reluctance of others to enter.

LETS boundaries are primarily normative, driven by a belief in and commitment to LETS as a valuable (complementary) means of providing and obtaining goods and services. As several writers have argued, the motivation to establish, and to join, LETS comes largely from cultural and ideological rather than labour-related concerns (see O'Doherty et al. 1999: 1643). Many members see LETS as a way of pursuing an environmentally sustainable, progressive, locally empowered way of life that challenges the power of larger, non-indigenous corporations. This does not mean that LETS members are consistently committed to a socialist or ecological politics; but while political and normative affiliations vary, research suggests that the pool of potential LETS members is structured by, at the very least, openness to a broad progressive ethic. Can this commitment protect and sustain LETS pathways from the centripedal pull described above as well as from the centrifugal effects of 'inefficient' transactions and policy disincentives, such as, in the United Kingdom, the threatened loss of welfare benefits as a result of LETS earnings?

A major difficulty in LETS mobilising normative commitment is that beliefs – for instance, in non-capitalist economic relations – are not, for the most part, collectivised or institutionalised, but held individually by LETS members. The emphasis of LETS schemes on member freedom also means that members do not need to adopt a particular political or ideological stance. Instead, many schemes articulate self-interest as a rationale for joining and participating. This is particularly evident in schemes, backed by community development workers, that target people on low incomes where the rationale for participating is identified in economic and social terms. More affluent, middle-class members may also see some self-interest in participating: access to particular kinds of goods and services (e.g., home-made produce) or developing new and stronger community ties, especially for recent newcomers to an area.

While some LETS schemes collectively, and members individually, have sought to mobilise a reconstituted notion of self-interest based on community, service and social bonds, the broader emphasis on self-interest proved to be a Trojan horse for many LETS. For, having established self-interest as a central motivation, if LETS cannot convince people on these terms, they will refrain from joining or else leave. Schemes deploying self-interest to create and maintain commitment therefore have two choices: either they show that LETS make good sense according to prevailing notions of self-interest, or they need effectively to redefine the self-interest to which they appeal. But leaving aside the general difficulties in recrafting something as embedded as self-interest, LETS' task is made harder by the very different interests members have. LETS want to argue that schemes are good for local people, yet the 'good' needs to be identified in a number of registers: from developing skills and confidence, to making friends, accessing affordable goods and services, and protecting independent local economies. While these may seem compatible at first glance, fractures begin to appear once they acquire a more concrete form.

Summerhill School

An American visitor . . . criticized our school on the grounds that it is an island, that it is not fitting into a community, and that it is not part of a larger social unit . . . Summerhill *is* an island. It has to be an island . . . We have many contacts with local people . . . Yet, fundamentally, we are not a part of the community.   (Neill 1968: 35–6)

In the twentieth century progressive schools came and went around the globe. One of the earliest, which was also to survive the range of assaults on it, was Summerhill. Located in a large house and grounds in Leiston, Suffolk, Summerhill is a small private school, educating sixty to a hundred children between the ages of five and sixteen. Summerhill School was established by the educationalist A. S. Neill in the early 1920s, according to principles of individual autonomy and community self-governance, principles which have remained at the heart of the school into the twenty-first century. What makes the school especially distinct, however, is the way in which it counters the subordination of children, and the pathways it makes available to them as a result (see, e.g.,

Gribble 1998: ch. 1). Summerhill provides an example of a community in which alternative pathways are rendered normal and beneficial by the school's wider, and very powerful, normative environment.

The guiding normative principle shaping both de facto and de jure pathways is that of freedom (Neill 1968: 108–11). Yet Summerhill's conception of freeedom is not a particularly unusual one; in many respects it converges with dominant liberal paradigms. It emphasises 'freedom from' rather than a more positive conception; and it distinguishes freedom from licence (Neill 1968: 105; see also Darling 1992). The school takes the latter distinction to mean that freedom to 'do' is curtailed at the point where actions detrimentally affect others (Neill 1968: 112). The main difference between Summerhill and wider liberal practices of freedom concerns the relationship between freedom and equality, especially of age. In Summerhill, all are entitled to freedom (except for adults, when it conflicts with their obligations), but no one's freedom takes precedence over that of anyone else or over that of the school as a whole.

The second core normative principle is participative democracy. However, it is important to emphasise the distinctive power of the headteacher, who was, at the time of my research, Zoe Readhead, Neill's daughter. As proprietor and head of the school, certain powers and decisions such as staffing, and health and safety are reserved to her (see also Appleton 2000: 108–10; Darling 1992: 51). More generally, the decisional realm of the school community is limited to those matters affecting the daily life of members. This community of children, teachers and pastoral staff is governed through a school meeting which, when I visited in 2002, met daily to make new rules (known as laws), hear grievances and act as a court in determining guilt and appropriate sanctions. Teachers are accountable to the meeting and, like the children, can be 'brought up' and fined. Laws emphasize the importance of coexistence in a community where little is compulsory or required. Rules, for instance surrounding bedtime, provide a way of managing conflicting interests and minimising 'nuisance' behaviour (see Appleton 2000: 96–7); other laws, such as around smoking and website access, seek to maintain a high level of personal freedom while 'protecting' younger pupils.

Within this normative environment, and the structures and rules it engenders, practices that would engender heavy costs in other schools – skipping class, 'messing about', 'cheeking' the teacher, or demanding rights – become reasonable, natural and 'affordable'. At Summerhill, children's ability to engage in the activities that suit them, even if it is skateboarding or tree-climbing for hours, their informal and peer-based, first-name relationships with adults, and their participation in making decisions, are seen as central to their necessary and rightful freedom. This does not mean that children always behave in preferred or desirable ways. Behaving badly, usually rendered legible in terms of the psychological 'baggage' children bring to Summerhill (Darling 1992: 47–8), can place a strain on the community. However, the school's capacity to fine-tune its practices, during the course of close to a century, has meant that the de facto pathways of children throw up few surprises. While they may not always follow the de jure paths established – of bedtime, meals, grievances, dispute resolution, conduct in town – they rarely, fundamentally, disrupt them.

Summerhill's capacity to facilitate counter-normative pathways depends not only on its structures and ethos but also on the ways in which it can protect them. Unlike the other communities discussed, with their informal, permeable membranes, Summerhill's educational status, ownership of land and income from fees gives it physically respected boundaries. These are not, however, impermeable; as a legal and physical entity, Summerhill possesses defined and regulated gateways through which state officials, children, school employees, parents and visitors gain varied access. At the same time, to see the wider environment being carried into Summerhill exclusively on the backs of the people who enter would be to underestimate the complex ways in which official laws, rules and values permeate. British law gives private schools a considerable degree of autonomy, an autonomy stretched to its full extent by Summerhill. Even so, through its long history the school has frequently rubbed up against the regulatory structure of the state in ways that threatened to undermine its distinctiveness, and even its existence.

Over the years, Summerhill has been repeatedly forced to accept a diverse and changing range of legal provisions. As proprietor and headteacher, Readhead has some capacity to determine which obligations become 'environmentalised' – that is subject to school interpretation and discretion – and which cannot (see chapter 7). Some of the most serious clashes with state forces, however, have occurred as a result of the school attempting to environmentalise requirements perceived by the state as non-negotiable. In 1999, for instance, conflicts with the Department for Education and Employment came to a head when, after a series of critical inspections, a report by the Office for Standards in Education (OFSTED) identified several major concerns.[8] In large part these related to the school's apparent disregard of wider institutional norms within the educational sector: 'The most serious difficulty for the school is that it does not agree that identified weaknesses in its provision are weaknesses: such judgements are seen as external impositions at odds with the school's belief and values' (para. 60). These weaknesses included children's lack of attendance at classes, the absence of required assessment and toilet facilities shared by boys, girls and adults. The inspectors' report led to a notice of complaint by the Department for Education and Employment. Summerhill was given six months in which to make the necessary changes. Failure to do so would cause the school to be struck off the register of independent schools and forced to close.

Readhead appealed against the notice of complaint. In the shadow of a court hearing, the matter was eventually settled, with both sides making concessions. Summerhill refused, however, to give in on the key issue of voluntary class attendance. As Readhead said, when interviewed on an American community radio show: there cannot be equal relations between children and adults if classes are compulsory. At the same time, the school formalised its curriculum planning, skills attainment and records procedures.[9] It also systematised its health and safety monitoring in the light of inspectors' comments.

A more complicated area concerned physical contact between adults and children. While a feature of the school was the relaxed, physical contact between teachers and children, Summerhill found itself, in a climate of intensified anxiety and solicitousness about child abuse, under formal and informal pressure to scrutinise such contact through a sexualised lens. For instance, the OFSTED inspection stated, 'some of the physical contact between staff and pupils could be misconstrued and as such is ill-advised' (para. 31). Summerhill's response demonstrates the difficulties of maintaining a determined normative stance, while recognising the risks to the school of poor publicity. Thus physical contact was buffered through a policy of preconditions and 'no-go' areas that simultaneously resisted and appeased wider public anxieties.

The example of physical contact demonstrates the extent to which Summerhill has not only had to contend with the interventions and pressures of officialdom. Concerns of parents and children have also left their mark. This was apparent in the drive for children to leave school with qualifications which, over the years, has shaped the educational provision Summerhill offered. Children and parents have also brought another aspect of the wider environment into the school – an aspect, in turn, reinforced by the media, and by the culture permeating the school, namely, principles of gender and sexuality. Loyal to Neill's (1968: 109) view that children should be free from being taught beliefs, Summerhill has eschewed equal opportunity policies and affirmative action. It sees the solution to social inequalities in making activities generally open, encouraging all children's self-confidence and using the grievance system to deal with harassment or prejudice. However, Summerhill's approach can be criticised for underestimating the power and pervasiveness of gender (and other) norms, a criticism – school members told me – often raised by visitors. It also assumes that legal structures, with their stress on an offender, a self-identified victim and a discrete remedy, can provide effective mechanisms for dealing with more systemic forms of harm and inequality. The limitations of this approach proved a live issue during my visits, when several staff and students referred to the routine, and largely unchallenged, use of 'gay' as a term of abuse.

In considering Summerhill, then, as a bounded community, we need to consider the external norms, values and relations that penetrate and structure different members' agency; the statutory and financial requirements that cut down agency; the centripedal processes, particularly in relation to qualifications, that make other choices sometimes appear to be more desirable and the centrifugal pressures of living in a small educational community with limited resources. However, against these forces, Summerhill draws upon two things to protect its practices and existence: first, the satisfaction of children and parents with the community and education provision offered (see Gribble 1998: ch. 1), and second, Summerhill's name, history and values. The school's clear, explicit and constant commitment to a set of founding beliefs is widely known and has generated, over the decades, considerable international support. The school has also achieved a high level of informed commitment from parents and students. In conversations with children at the school, I was struck not only by their loyalty to the school but also by their understanding of its core motivating values of freedom and democracy.

The potential of permeability

So far I have discussed the problem of permeability, in relation to Greenham, LETS and Summerhill, as one of the outside seeping in. In this final section I want briefly to reverse the terms of discussion to consider the wider impact that bounded communities can have. Separatist communities have long been criticised for privileging an idealised micro-society over the pursuit of wider change. While I would suggest that even apparently closed communities frequently impact beyond their boundaries, this is more obviously the case for the kinds of prefigurative communities discussed here, since the porosity of their perimeters enables two-way movement.

In his discussion of the Pollok Free State, established in 1994 to protest against the building of the M77 motorway in Scotland, Seel (1997) usefully explores the wider impact of protest communities. His argument centres on the capacity of such communities to provide a site for intensive action and as a forum for building alliances, to politicise local people, operationalise counter-hegemonic values and generate wider media debate. Seel's (1997) analysis echoes the findings of other writers who have explored protest communities. Greenham peace camp, for instance, played a significant role within wider peace and feminist politics. Camp women spoke at meetings at home and abroad, fronted a legal challenge in the US courts to the deployment of cruise missiles in England and played a high-profile role in a range of protests, from women in prison to the arms trade, vivisection, uranium mining and the British miners' strike in order 'to draw attention to the connections between nuclear militarism and other issues' (Roseneil 1995: 113; see also Finch et al. 1986). Greenham's role in the peace movement generally, as well as the camp's involvement in wider activities, drew mixed reports. Yet both the praise and resentment were products of the amplified capacity of communities such as the women's peace camp to 'represent spaces from which individuals can go out and work and play in external political spaces' (Sargisson 2000: 72).

Greenham's wider contribution cannot be reduced to the explicit political activity its members pursued beyond the camp's boundaries. As well as changing the habitus of people who visited, and in particular lived at, Greenham, the camp also operated as a showcase for a different way of life, oriented around ecological, communal and democratic practices (Roseneil 1995, 2000; cf. Emberley and Landry 1989; see generally Pepper 1991: 184–9). One aspect of this showcasing to which I want to draw attention relates to the blurring of distinctions between the intimate and impersonal, between the spaces where public or private norms prevailed. In the case of Greenham (and similar communities), outsiders were welcomed into a space that was both home and not-home. Strangers entered and shared the tangle of domestic life – that is both the strangers who coexisted within the community as well as those who peeped in during the course of a temporary visit. As a result, aspects of life, normally withdrawn from wider scrutiny, were played out not only differently, but with much greater visibility. Summerhill today offers a similar kind of showcasing. Visitors are welcomed to the school, although in ways circumscribed according to time, space and activity. When I attended, a sign near the front door indicated that visitors were restricted from entering bedrooms, staying into the evening, and from using the swimming pool or Climbing Tree.

Living differently may impact on participants and visitors, but central to my argument is the material effects of counter-normative practices. These effects cannot be restricted to explicit forms of political pressure or instrumental participation within local democratic decision-making. They also do not just relate to government. The plethora of Greenham imagery, representations and talk, during the early to mid-1980s, demonstrate the ways in which alternative communities can 'adulterate' the cultural landscape, at least for a short while (see Emberley and Landry 1989). Yet, as the Greenham example highlights, the effects of alternative communities may not be to move wider society 'forward'. Accommodation or domestication of the cultural other may take place almost entirely on the terms of dominant values. Summerhill's experience with OFSTED also demonstrates how official forces can choose not to accommodate alternative ways of doing things and may seek their elimination instead. A more uneasy, less overtly hostile, form of coexistence occurred in relation to LETS. While some governments, Australia being one example, attempted to accommodate LETS schemes, the British government adopted a more equivocal approach. Although local government in some districts helped fund LETS schemes as part of their community development work, national government proved less supportive, paradoxically through their recognition of LETS currency. In other words, their treatment of LETS credit as 'real' income required, controversially, that it be declared for social security purposes.

How governments respond to community practices is a live issue that may, in part, be shaped by social protest. Although this did not prove to be successful in relation to LETS and welfare benefits in the United Kingdom, one context in which social activism drove governments' response to new social forms was in relation to lesbian and gay partnership recognition. As I discussed in chapter 5, the pursuit of lesbian and gay marriage was, in large part, premised on the demand that the state recognise and respond to changing family forms – that it shape itself more ergonomically and proximately around new social structures. While studies in governmentality emphasise the extent to which governmental techniques shape social identity and conduct through a set of linkages that allow state bodies to act at a distance, my argument here is predicated on the reverse set of effects: that changes routinised in social conduct, allied to political activity, affect the governmental techniques deployed and the programmes with which they are associated.

Whether changes in social conduct have these consequences depends on several things. These include the 'rub' factor – namely, those communities with more permeable boundaries may impress more strongly on wider social relations, even as their limited autonomy restricts their capacity to sustain counter-normative practices over time. It also depends on the configurations of social forces operating at those junctures where community norms and practices meet wider society. For instance, the inspections of Summerhill School provided such a juncture. While inspectors could, and did on some occasions, play a more mediating role, adapting their inspection in recognition of the values of Summerhill School, where this did not occur, as in the late 1990s, state and school came into collision. I have described above the range of forces that mobilised at this juncture, and the school's (temporary) victory, powerfully symbolised by the impromptu meeting held at the High Court to decide whether the government's offer should be accepted. As Appleton (2000: 266), an ex-Summerhill teacher, described, 'There really is something extraordinary about this case in which High Court proceedings are stopped and government officials are forced to wait, while a meeting of children decide whether or not to accept their proposals.' However, what did not happen in the 1990s was the permeation of Summerhill principles into British education more generally. While the conflict may have provided the space for this to occur, the limited involvement of other educational institutions allowed the government to accommodate Summerhill without having to extend, or respond to, progressive education more generally.

Conclusion

In this chapter I have explored attempts to develop and sustain collective counter-normative practices. While the case of Greenham illustrates a community developed in conditions of protest, my focus here has mainly been on the other ways in which counter-normative communities challenge the status quo. Protest-based communities can be seen as contesting the extremities or terminal forms of rule – road building and military bases, for instance – but it is when communities establish, and are able to sustain, new forms of social practice that they provide an alternative to more conventional ways of doing things. This chapter has outlined the counter-normative aspects of the three communities studied, particularly in relation to norms of freedom, democracy and community, and to organising principles of gender, class, sexuality and age. As I suggested at the outset, one aim of this chapter was to explore the relationship between equality and diversity within progressive communities, to see whether their intersection provides evidence for the claim that equality and community are the enemy of freedom and diversity.

The three communities, in their different approach to equality, highlight the limitations of drawing abstract, universal equivalences between community and sameness. While none of the communities explicitly sought to defend existing asymmetries, the focus of Greenham, LETS and Summerhill were on particular inequalities, namely of gender/sexuality, wealth and age respectively. Likewise, their strategies for pursuing equality ranged from Summerhill's formal (though partial) elimination of age-based inequalities of power, to the revaluations of goods and services in some LETS, and the more informal cultural norms and ousting of men at Greenham. At the same time the three communities shared similarities in their focus on the individual as equality's subject, and in their emphasis, despite the stress each placed on spontaneity and self-governance, on treating formal status as equality's 'what'. In other words, none of the communities tackled to any considerable extent the informal ways in which differences of power and status were maintained.

In each of the communities formal equality was also coupled with a premium placed on freedom and diversity. At Greenham it related to each participant's right to choose her particular contribution to camp life and protest activities. LETS emphasised freedom to trade and the need for diversity to make the scheme work. At Summerhill equal individual autonomy has the highest formal status of any of the three communities. In contrast to Greenham, the school has also adopted a form of decision-making which seeks to encourage rather than erase difference. Greenham was not as wedded to consensual decision-making as Seneca peace camp, and frequently used a more anarchic 'hundred flowers bloom' approach to avoid conflict, however it tended to seek agreement as a way of acknowledging women's different concerns. Summerhill, in contrast, has a majority-based approach to decision-making. Appleton (2000: 106) argues that this approach works better because it means that people do not feel obliged to agree, and there is less drive to iron out differences of opinion. Indeed, outvoted individuals can, if they wish, repeatedly bring an issue back to subsequent meetings until they get the outcome sought.

Each of the communities then, albeit in differing ways, reveals a commitment to, and capacity to support, equality, freedom and diversity, as they also demonstrate the limits on each. These limits, as I discussed in earlier chapters, are inevitable in any community or society, since it is impossible to avoid exercising a steer in favour of or against certain practices or beliefs. But this steer can be much less directive; it can take a far more environmental form to structure rather than eliminate agency than many critics of community suggest. In the remainder of this conclusion I want to move away from the content of community norms and practices to the role played by boundaries in maintaining an environment within which counter-normative practices can thrive. The creation of a conducive environment is central to shaping and sustaining alternative pathways. This is clearly illustrated in the case of ecologically sound practices (e.g., Sargisson 2000: 64–5). While some ecological practices, such as environmentally friendly sewerage, *depend* on institutional or community-level reforms to come into existence, others, such as veganism and recycling, are simply far easier to maintain in a collective setting which treats such practices as beneficial, normal, and aesthetically pleasing, while rendering them relatively cost-free in terms of time, effort and other resources.

Maintaining conditions that contravene or clash with wider environmental norms and rationalities depends on buffering processes that separate and protect them. A key theme in this chapter has been the difficulties communities encounter in their attempts to establish effective boundaries or membranes. But is porousness or permeability necessarily a bad thing? The chapter opened with the argument that boundaries are necessary to protect counter-normative community practices. I want to close by examining the benefits that come from permeability.

Whatever relationship a group establishes with its environment, with the world outside its borders; however well it patrols its boundaries; and no matter what it decides to accept or reject of the life of the outside – this relationship is subject to continual revision in the face of changes in that external environment.   (Kanter 1972: 144)

Each of the communities in this chapter demonstrate the different and varying ways in which wider society intrudes upon the crafting of community norms, institutions and practices. Not always on the backs of new members, norms, rules, laws, processes, ideologies and institutions shape community life in accommodatory and resistant ways; people are enticed to leave, those who stay face the costs of doing so, and isomorphic practices, symbolically replicating the 'outside', evolve. Undoubtedly this may limit the counter-normative pathways communities develop. At the same time, it is what enables communities to communicate and be intelligible beyond their own boundaries, to showcase their pathways and institutional structures, and in the process recruit new members. However, a further benefit of permeability is in the way in which communities impact on life beyond their borders. While writers have explored the symbolic, practical and strategic forms this can take, one less addressed form, discussed in this chapter, is that of recognition. This is a question not just of *whether* wider political and cultural forces recognise community practice, but *how* they do so. I mentioned above how LETS credits were treated as income for welfare benefit purposes; Summerhill similarly experienced the equivocal status of recognition. As a private school, Summerhill is officially recognised by the government and integrated within its official systems, but as a system of education, Summerhill felt that the Department for Education and Employment had no understanding or interest in what it offered. Recognition is important to the processes by which community practices impact more widely. While recognition can be favourable or unfavourable, and while a refusal to recognise does not necessarily mean a lack of impact, processes of recognition engage with the question of what a community is about, what relevance it has for others, its broader contribution and the wider challenges it generates.

The importance of recognition places a premium on boundary actors who straddle the space where community and wider society meet. This may be a physical space, but it can also be a normative or symbolic one. The synthesis of the two can be seen in the activities of Greenham members, who brought Greenham into the mainstream through their court protests, media work and public speaking. However, in many respects, all members of the communities discussed here were boundary actors. In part this took a spatial form of entering and exiting, but it also occured through the isomorphism in which all members participated, particularly evident in LETS, with its talk of currency, accounts and exchange, and in the juridical orientation of Summerhill with its laws, fines, dispute and grievance structures. While this isomorphism can be read as the penetration of wider society, it also provides a common language through which community members can make sense of their processes and practices both to themselves and to others, reducing the need to translate community institutions into wider public discourse in order to become intelligible.

Permeability, intersection, hybridity and isomorphism express the meeting place between many prefigurative communities and wider society. But if these junctures and processes are crucial to the diffusion of new practices, are community boundaries needed in the first place? Do such protectionist attempts, in their alienation of wider society, in fact undermine the ability of communities to communicate and impact further afield? I want to argue that they do not. Boundaries – whether legal, spatial or normative – are necessary for new pathways and social environments to develop. And boundaries are necessary if community members are to straddle or cross them. We can see this process as constituting a double counter-hegemonic and anti-hegemonic move. In other words, the boundary provides a way of enabling new normative and social practices and institutions to develop within a community; it also allows members to disseminate new norms and practices more broadly. At the same time, boundaries make a twofold critique possible – that is a critique of wider social practices as well as a critique of community ones.

Diversity politics provides a discursive space in which both anti-hegemonic and counter-hegemonic processes are advocated. However, writers have tended to valorise one side of the equation at the expense of the other. My argument in this and previous work has been for the need for both: for the creation of new ways of doing things anchored in and driving new normative foundations, at the same time as critical and deconstructive perspectives maintain a reflexive approach. The porous boundaries engendered by alternative communities make both movements possible. Yet questions still remain. In particular, I am interested in two. First, what norms provide the basis for internal critique? In other words, is critique driven by norms and values prevalent within wider society or by those advocated within the new community? Second, how do alternative community practices relate to wider pockets of support or shared values? My premise in this chapter has not been that communities face a pervasively hostile wider environment. Given the contradictions and heterogeneity in the wider environment, norms and institutions conducive to alternative practices do exist. Working with this dilemmatic environment was the challenge identified in the previous chapter. How it relates to the boundary process and the establishment of alternative communities is one of the tasks of the final chapter that follows.

Notes

1. A. S. Neill (1968: 23).
2. According to Sasha Roseneil (2000), Greenham was not just a peace camp with a tangible location, it also existed as a network of individuals and groups. Kirk (1989: 270) makes a similar point about the early years when 'Greenham was as much a state of mind as a place'. As this chapter concerns communities as spaces of living, I will focus on the camp – the physical 'hub' of the Greenham network, as this is where alternative social pathways were most apparent.

3. This was captured in the documentary *Carry Greenham Home* (1983), by Beeban Kidron and Amanda Richardson, distributed by Contemporary Films.
4. There has been some debate as to whether LETS currency is real money, since it does not have a physical existence and only emerges through the recording of transactions undertaken; see Bowring (1998: 86), Dobson (1993), Pacione (1997: 1186).
5. LETS have been described as a particularly effective way of facilitating social ties – providing a more material basis (than coffee mornings) through which relations and connections between people can develop.
6. A similar tension has been identified in research on communes (see Abrams and McCulloch 1976: 180), where there is also often an unmet need for a range of relevant skills to enable the commune to function as a richly complex network of interdependence where 'what each does fits into the miniature economy as a whole'.
7. Croall (1997: 65) suggests that firms joining LETS are particularly concerned with the reliability of the services offered.
8. See OFSTED Report on Summerhill School, Inspection, 1–5 March 1999, Reference 144/99/IND.
9. http://www.alternative-learning.org/ale/sum-faq.html.

9

Diversity through equality

Challenging Diversity offers an intervention in a set of debates carved out across and through that place where the politics of equality meets the politics of diversity. In the process, my argument has focused on two sets of issues. The first relates to diversity's boundaries; on what basis can distinctions be drawn between those differences to be encouraged and enabled and those to be discouraged and undone? The second set of issues concerns the wider normative context within which struggles for equality and diversity are placed. This normative context can be seen in two quite different lights. As a conservative configuration of principles, it can work to 'hold back' the pursuit of change. At the same time, progressive norms provide a thicker, richer texture to the question of what diversity through equality actually entails.

My starting point has been the celebration of diversity – the valorisation of social and, in particular, cultural differences by those on the poststructuralist left. Yet, as I explored in the book's early chapters, despite the praise heaped upon a flourishing pluralism, few diversity theorists have been prepared to discard completely notions of value. As a result, the claim to celebrate diversity quickly boiled down to a debate about where the limits of diversity should lie. Addressing this issue has been complicated by the fact that diversity politics recognises two different kinds of diversity as troublesome: differences of power, organised in particular around gender, race and sexuality, and social and cultural practices perceived as harmful or undesirable. These two categories are not discrete, separate entities; cultural practices, for instance, may express gendered or sexualised inequalities; nevertheless, they look in different ways. While opposing social inequality critically focuses on the structures, institutions and common sense of mainstream society, the problematic of cultural harm faces towards the distinctive practices of minority communities.

One danger, in my view, for diversity politics lies in the increasingly dominant position achieved by this second approach of cultural harm. A reason for this may be the discursive influence of liberal voices in the field. While these voices do not uncritically confirm the status quo, their vision is trained on evaluating that which is different. This tendency can be seen in the work of Okin (1998), who, while recognising that majority cultures can be more patriarchal, for instance, than minority ones, still focuses her critical attention on the gendered character of the latter. Galeotti (2002) similarly, in a tightly argued book on the politics of diversity, largely directs her remarks to the problematic of minority practices. This reorientation of attention is also apparent in more radical writing in the field. While authors such as Fraser (1997: 185–6) and Mouffe (2000: 20) explore diversity's limits within a wider analysis of hegemonic practices and structures, the reorientation of the field towards cultural harms facilitates the detachment of diversity's evaluation from this wider analysis. As a consequence, exploring diversity's limits tends to flag up the distinctive rather than the commonplace, practices, groups and identities rather than institutions and systems, and the conduct of minority, rather than majority, constituencies.

Recentring inequality

In engaging with this debate, my approach has been somewhat different. More particularly I have sought to recentre social inequality, rather than cultural harm, as diversity's problematic, a recentring developed over the course of the book, although set out in most detail in chapter 3. The driver in developing this approach has been a set of dilemmas in equality theorising. These revolve around the following questions. Are conduct-based discrimination and the asymmetrical effects of norms inevitable? What is the relationship between conduct and identity, especially when benign groups do 'bad' things? And how can equality politics turn its attention towards that which is least visible and most embedded?

In chapters 2 to 4 I explore equality through the lens of diversity. I want to summarise my argument here because it provides an important conceptual, analytical and normative foundation for the book as a whole. Broadly, my argument has four main elements. First, I argue that societal and governmental differentiation between practices, values and preferences is unavoidable. While such differentiation might, theoretically, involve neither legal proscription nor permission, a 'morally and culturally neutral state . . . is logically impossible' (Parekh 2000: 201–2). The implications of this are important. It does not mean, as some might think, that social inequality is unavoidable, but it does mean that certain choices, preferences, activities and values will be produced, preferred and enabled over others. As I set out in some detail in chapter 3, what follows from this is the need to distinguish between inequalities that should be eliminated, and social policy choices that invariably emerge within a society in one form or another. The judgment inherent in these choices is not a bad thing. What matters is how they are made, the values underpinning them, and the effects of their operation. Without suggesting that societies are sealed-off entities, capable of living in radically different ways from their neighbours, and unaffected by global forces, the choices made within them are crucial to the social politics pursued.

My second foundation, explored in chapter 4, concerns the complex connections, but also the elasticity and contingency of the links, between beingness, identity and conduct. Rather than seeing them as separate – dividing public from private selves and conduct from both – left diversity politics has emphasised both the fluidity of identity (Weeks 1993: 202–3) and its tangled relationship to other statuses and practices. These connections are important to thinking about equality. If the way in which practices are treated affects people's sense of self, an argument can be made that all conduct or preference choices need to be equally applauded if people are to be treated equally. However, this stance not only conflicts with my claim above that evaluative policy choices are necessary and inevitable, it also conflicts with a notion of substantive equality that is predicated upon undoing or eliminating practices of domination. The approach to this dilemma that I have adopted looks both ways. It seeks to recognise the investment people may have in particular conduct, especially when it is under attack, as in the case of cigarette smoking. At the same time, this cannot be used to evacuate the field of policy judgement. Recognising the elasticity of the relationship between identity and conduct, as well as its contingency and unpredictability (Halley 1993; Stychin 1996), makes interventions possible. The same is true for the relationship of identity to beingness. Here too the political stakes change if being is not collapsed into identity. This does not mean that a shell of beingness exists filled by the meat of identity. But it does mean that identities can be engaged with through institutional and other practices, without this inevitably undermining the equality of beings.

My third analytical foundation concerns the social importance of taken-for-granted, routinised practices and commonsense norms and relations. Clearly, what these are varies with the society in question. For instance, in certain polities, female veiling is routinised and unquestioned. However, in a country such as Britain, gendered minority practices receive a level of critical attention out of all proportion to their pervasiveness or influence. Aside from the ways in which this can reinforce cultural prejudices and social marginalisation, it also deflects attention from more embedded mainstream practices. While I agree with Nancy Fraser (1997: 187) that 'we should develop an alternative version that permits us to make normative judgments about the value of different differences by interrogating their relation to inequality', the challenge for diversity politics is also to look away from that which stands out as different in order to be able to evaluate the mainstream, the common and the 'normal'. This links to my final foundation: the institutionalised, systemic processes through which differences arise and are made to speak. In other words, rather than placing practices, groups and identities at the analytical and critical centre of debate, I argue for the need for a more structural approach.

In chapter 4 I set out my approach to equality, treating the individual as equality's 'who', power as equality's 'what' and dismantling structured relations as equality's 'how'. The cornerstone of this approach, in the light of the four foundations identified above, is the role played by organising principles of inequality. But what are organising principles of inequality? How are they to be defined and where are their limits to be drawn? In chapter 3 I explore these questions at some length, my aim being to elaborate a way of thinking about inequality where it does not collapse into social value. For the danger of this latter is the devalorisation of those constituencies whose conduct diverges most dramatically from dominant social norms. In other words, echoing arguments made by Catharine MacKinnon, those who are most marginalised or disadvantaged are likely to be hardest placed to come within equality-based frameworks if these frameworks depend on demonstrating common or congruous qualities with dominant and privileged constituencies. This does not mean that norms and evaluation are irrelevant. Simply, they cannot be trusted as the basis for making judgments in cases where social inequality comes into play. For the operation of social inequalities will also shape the norms and values brought to bear in evaluating particular practices, interests and desires.

My exploration of the distinction between principles of inequality and other forms of discrimination drew on the example of smoking – a practice, I argue, that could legitimately be disadvantaged for policy reasons, since it did not constitute a principle of inequality. It therefore differs from an inequality such as sexual orientation. Yet questions about the hetero/homo divide have also tended until very recently to be addressed in public policy terms. Debates about how to respond and how to regulate have revolved around whether sexuality is chosen or immutable, whether as a minority form of conduct it is harmful or merely offensive, and the implications these determinations have for questions of agency and responsibility. My emphasis, by contrast, on whether disadvantage operates as a structural form of social inequality, which I treat as found in the case of sexuality within modern Western societies, brackets, at least to begin with, the question of harm. This does not mean that harm is irrelevant in all circumstances, on the contrary. However, since what counts as harm depends on the values underpinning it, harm does not provide a secure basis for determining whether structural forms of disadvantage are or are not acceptable.

In chapter 3 I set out the main requirements for a social asymmetry to constitute an organising principle of inequality. My argument there is that while the unequal treatment of people based on their conduct, beliefs or social location is a central and necessary element, it is not sufficient. Two other elements are also crucial. The first is the capacity of the inequality to shape other dimensions of the social. This includes the capacity of dominant interests to permeate and circulate through other practices and institutions. However, equally important is the way in which norms, values, truths and even epistemological frameworks articulated to particular inequalities prove socially pervasive. This pervasiveness is evident in relation to principles of race, socioeconomic class and gender. A case has also been made that the values, norms and standards associated with the hetero/homo divide also pervade social life and current epistemologies (Sedgwick 1990). However, there is far less evidence that this is true for smoking. This distinction is important, since the reason for differentiating inequalities from other forms of social disadvantage is that we cannot trust ourselves to judge in a context where the conditions, processes and criteria for judging are shaped by, for example, gendered or racialised cultural and epistemological standards.

The second additional element that I claim is necessary for disadvantage to constitute an organising principle of inequality follows from this. It concerns the ways in which principles of inequality significantly impact on social dynamics such as the intimate/impersonal, capitalism and community boundary maintenance. In the case of sexuality, the hetero/homo divide can be seen as playing a central role in shaping the social dynamics of desire and, to a lesser extent, the intimate/impersonal in modern Western societies. This mutually constitutive relationship gives principles of inequality their strength and durability as they shape, and are shaped by, central and enduring aspects of society (even as both also evolve and adapt). Other forms of conduct-based disadvantage may be generated or sustained by particular social dynamics. However, their inability substantially to structure these dynamics means that the latter retain some autonomy. In other words, to take the cigarette example, discrimination towards smokers may be shaped, in complex ways, by the dynamics of capitalism or community boundary formation, but these dynamics are not sustained or given shape, in any significant way, by inequalities forged through smoking. Cigarette smoking does not organise the social and, as a consequence, it does not act as the lever in maintaining its own unequal reproduction. Social dynamics, with a high level of independence from smoking, can, instead, organise the social place of smoking in rapidly changing and flexible ways.

Yet, as I stress in chapter 3, the boundary between social differentiation and social inequality is not rigid and unchanging. Structured forms of disadvantage may implode into multiple fragments which can be less easily harnessed to the logic of a binary inequality. Alternatively, inchoate forms of disadvantage may become consolidated and emerge over time as principles of inequality in their own right. But even if differentiations do not have the status of social inequalities, this does not mean that discrimination is necessarily acceptable or benign. Aside from the important ways in which conduct-based differentiations, such as smoking, may be linked to organising principles of class or gender, questions of proportionality come into play, in part to head off the process of turning a legitimate policy-driven disadvantage into a principle of inequality.

Beyond sameness

Where principles of inequality are identified, my argument, developed in chapter 4, is that these should be dismantled. But what implications does such a dismantling, in pursuit of greater equality in the individual's exercise of power, have for diversity? Many voices within the field of diversity politics, particularly those influenced by poststructuralism, have argued in recent years against affirming equality on the grounds that it is grounded in sameness (e.g., Flax 1992). In a tightly argued article Cynthia Ward (1997: 98) argues that '*any* acceptable justification of equality requires the establishment of some descriptive sameness among people'. Ward (1997) argues that some common quality of people is necessary as a justification for equality, but she presumes rather than explains why the difference default is inequality. I want to leave to one side the question of what equality requires in justification, since, in my view, its justification lies in irreducible norms and values rather than social facts. My premise is that equality functions as a norm, strategy and aspiration. As such, it does not entail sameness. Nor does it have to be interpreted arithmetically. It is unsurprising that equality has been read as such, given its associated terminology of 'more' and 'less'. However, once equality is treated as a normative project involving irreducibly different phenomena, the notion of sameness becomes redundant. The incommensurability of equality claims, and the need to go beyond arithmetic politics, can be seen in demands for equality of recognition. Here, not only the constituencies and practices to be recognised, but the nature of recognition also, involve qualitative differences. While attempts to achieve proportional representation function as a way of creating a concretised and quantifiable form of recognition, cultural forms of recognition cannot be reduced to mere quanta. This is also true, if not more so, in relation to the broader concept of equality of power. It is meaningless to reduce power to an amount; and equality does not necessarily require that the same kind of power be exercised.

However, my rejection of the assertion that equality is inherently about sameness goes beyond the conceptual issue of quantification. More substantively, my argument is premised on the claim that many benign forms of diversity depend on greater social equality to come into existence. This premise was, as I have said, the early basis for diversity and multicultural politics, as well as for more radical theoretical perspectives, concerned for the ways in which inequality undermined particular cultural and social forms. Writers such as Kymlicka (1995) have explored the value of collective and individual rights in enabling ethnic and indigenous constituencies to continue and flourish; my focus, in this book, has been on the relationship between (in)equality and a more social politics. In particular, the previous chapter flagged up the contribution of external economic resources and internal democratic modes to the survival of the alternative communities discussed. Many Greenham and LETS members, for instance, depended on social security and welfare benefits, and would have been compromised in their ability to continue as members without them. Likewise, Summerhill's ability to support diversity among its young members is contingent on the school's practical commitment to equality and democracy.

The experience of Summerhill, Greenham and LETS casts doubt on the double-headed attack on equality and community, both of which are seen to deny and repress difference (Flax 1992: 194; Young 1990a: 227). While it is the case that some communities stress conformity and uniformity (see, e.g., Kanter 1972: 18), in many diversity is emphasised. It may seem unfair to criticise Young on this point, as her discussion is largely about ideal and abstract rather than actual intentional communities. However, it is easy to move from this theoretical claim to one which sees concrete communities, particularly equality-based ones, as threatening diversity and freedom. This presumed antagonism between equality and autonomy sidelines the very important ways in which equality – or positive freedom – enables diverse ways of being to flourish. However, this does not mean that diversity can expand indefinitely. As my discussion of the eruv, in chapter 2, highlighted, enabling and supporting certain cultural or social practices may, in turn, impact negatively on others. But, as with equality, a quantitative approach to diversity is the wrong approach. The important issue is not how much diversity – even assuming that this could be measured – but what kinds of diversity do different social environments produce?

Challenging Diversity has addressed this issue in several respects. In chapter 7 I drew on the concept of the dilemmatic environment to explore the ways in which similarly situated people may make different choices. The dilemmatic environment is one in which norms, values, resource allocations, power relations and institutions converge to create an ambivalent logic (cf. Honig 1996). In other words, choices are not straightforward or clear-cut. Whatever decision individuals make, they feel the tug of their alternatives. Below, I consider dilemmatic environments further in relation to individual conduct within prefigurative communities. However, here I want to consider the question of the forms of collective difference that can coexist with, and are produced out of, the undoing of social inequality.

In chapter 4 I argue for the dismantling of principles of inequality. But what does this mean for the social and cultural differences with which they are associated? Does a normative stance of radically dismantling inequality risk eradicating benign and valuable, as well as malign and worthless, differences? Does it take away the possibility of social forms of collective distinction, as a result reducing diversity to the level of individual conduct limited only by the boundaries of intelligibility within a given community or society? In asking these questions, I want to bracket two considerations: the emergence of new inequalities as older ones perish and the implications of global relations and structures for the undoing of inequality within a society. While crucially important to equality in practice, they can be sidelined for the purposes of this discussion, which focuses on the normative imagination underpinning the drive to dismantle relations such as gender, race and sexuality.

The claim that social inequalities should be undone, that is, converted into meaningless distinctions, needs complicating. Although I have treated gender, race, class, bodily capacity, sexuality and age as organising principles of inequality, they are not identical. Liberal multicultural theorists have tended to divide them according to whether they constitute thick, discrete cultural entities such as ethnicity, or immutable intra-cultural bases for unjust social distinction (Kymlicka 1989: 165, 1995: 19; Parekh 2000: 3–4). More radical scholars, such as Nancy Fraser (1997), have focused on the economic and cultural underpinnings of different inequalities and the kinds of justice to which they consequently give rise. While Fraser argues that justices of recognition and redistribution can involve both a reform-oriented and deconstructive/transformative mode, her approach tends to contrast a class politics, which seeks its own elimination, with a politics of sexuality, oriented towards the survival, recognition and valorisation of its minority forms. However, while useful and insightful, as this book has made clear, neither of these approaches quite accords with the one I wish to adopt. In earlier chapters, I mapped one approach to differentiating between principles of inequality based on their relationship to particular social dynamics and norms. What I want to consider here, however, relates to how various differences are normatively imagined.

Within the context of Western developed societies, at the turn of the twenty-first century we can envision a radical politics that divides inequalities into three categories: those to be eradicated, such as socioeconomic class; those whose meaning and effects need reassessing, such as age; and those which should become insignificant forms of difference, such as gender. However, this division brings immediate questions and concerns in its wake. One, unsurprisingly, concerns the particular allocation of inequalities among these three categories (cf. Fraser 1997: 200–2). Another is the notion that social categories should be eliminated or rendered meaningless. Does a strategy of dismantling also discard the normative, epistemological and cultural benefits of particular, inhabited, social locations?

In chapter 4 I raise the question of whether specific cultural differences can outlive, without refortifying, the inequalities that produce them and give them meaning. There I discuss one social movement to have developed this claim in a way that went beyond a politics of pride. Transgender activists have creatively argued for gender as a form of stylisation or modality through which social interaction, desire and preferences are articulated. While this argument is limited as a commentary on the present by its inadequate recognition of the power relations generated by gender, particularly the male/female binary, it does provide a vision of potentiality that can be integrated within a radical equality project. To abolish gender, from this trans perspective, would remove a language or modality from the repertoire of cultural and social possibility. While for some transgender activists what is at stake is the possibility of living out a coherent, binarised gender, for others, it is the creative and productive hybridity made possible by combining gender's language into new configurations.

Retaining the variety of tastes, manners, preferences and desires associated with gender, disarticulated from any particular sex-embodied form, may seem a relatively innocuous mode of diversity, particularly if cultural variations oriented around superiority and inferiority are excluded (Fraser 1997: 203–4). But is this the only form of diversity to which principles of socioeconomic class, gender, sexuality, ethnicity, age and race can legitimately give rise? And what is to happen to forms of variation that do not conform to these strictures? Clearly, views among radicals about the normative extent and character of social diversity vary. Jeffrey Weeks (1993: 208), for instance, adopts a far more pluralist approach than Nancy Fraser when he states, 'one should be free to choose between different values, different identities and communities', although how this intersects differences currently organised as principles of inequality remains unclear. Chantal Mouffe (2000: 21–22), similarly, recognises the constitutively necessary character of pluralism when she writes, 'The democratic society cannot be conceived any more as a society that would have realized the dream of a perfect harmony in social relations. . . . The main question of democratic politics becomes then not how to eliminate power, but how to constitute forms of power which are compatible with democratic values.'

At the same time, a major theme running throughout Mouffe's work is the need for a counter-hegemonic articulation among progressive struggles based on a radical democratic interpretation of liberty and equality. This articulation, and its implications for other antagonistic struggles, highlights the need for social relations themselves to change also. But it is this question of technologies of change which raises particularly difficult questions for a radical diversity politics. For, while democratic forms of self-governance, as illustrated by my discussion of communities in chapter 7, allow freedom and the elimination of domination to dovetail, other modes of change evoke the spectre of strategies of domination being deployed to eliminate injustices of power. This issue is far larger than I have space for. However, I raise it because the role of law and governance – formal and informal – has been sidelined in much radical theorising within the space of diversity politics which delineates what ought and ought not to happen without quite explaining the means of bringing these transformations into being. As a result, practices, identities and relations can be read as problematic or non-egalitarian, that is as things which should ideally not exist (e.g., Fraser 1997: 203–4). But does this mean that they should be prohibited, discouraged and avoided, or should they be tolerated at the margins? Because radical normative writing in the field of diversity politics has tended not to address the broad institutional implications of its claims, these differences of response are not teased out.

Galeotti's (2002) work may be useful here. Galeotti (2002: 42) argues for a threefold categorisation which encompasses differences that command respect and should be accommodated, those that are ethically incompatible with, and threatening to, the normative and social order, and those that involve the cultural subordination of group members. Galeotti argues for the elimination of the second group. However, she claims that the third category, since it does not threaten the normative and social order as a whole or cause physical harm, and since attempts to eliminate such differences through coercion usually prove costly and ineffective, should be tolerated. Galeotti's work is very much embedded within a tradition concerned with the survival and flourishing of liberal society; as such her arguments cannot be appropriated wholesale in the service of a more radical account. Nevertheless, her typology is useful because it reminds us that social formations can flourish and survive even if, and sometimes because, they embrace non-normative practices, institutions and preferences.

The capacity to incorporate practices, institutions and preferences which do not conform to prevailing norms is particularly evident when such practices or institutions are minority ones. At the same time, struggling with unwanted practices within the contours of normative theory raises the question as to why diversity is of value. To allow only for differences that are fully compatible with a normatively envisaged future suggests that the only benefit diversity brings forth is that of individual choice and social colour. But as Parekh (2000: 167) also argues, 'other cultures . . . enable [people] . . . to view their own from the outside, tease out its strengths and weaknesses, and deepen their self-consciousness'. Parekh's words are premised on a model of thick, bounded, ethnically based diversity. Nevertheless, his comments are instructive in relation to the imagined value of non-normative practices and values within a counter-normative social project. In other words, for a project that seeks to avoid entrenching new conservatisms, the margins provide both a space and a momentum for anti-hegemonic as well as counter-hegemonic challenges. I return to this issue below in relation to alternative communities.

Mobilising at the normative interface

Intersecting the issue of diversity's limits – and specially flagged up by the project of imagining a more equal society – the role played by other social norms and values forms the second cluster of issues explored in *Challenging Diversity*. My interest configures around several questions. Do dominant normative principles undermine or compromise the pursuit of equality? To what extent do equality reforms compel normative principles to change? How are particular versions of the diversity/equality relationship rendered normal and commonsensical? And what contribution do prefigurative communities make, taking into account the permeability of their boundaries, to the pursuit of a counter-normative settlement?

The first of these four questions proved the focal point for my discussion in chapter 5. There I examine the campaigns and law reform taking place, in a number of countries, to extend spousal recognition to same-sex couples. These reforms have been presented and rationalised in terms of equality, namely, the right of couples, regardless of their heterosexual or homosexual status, to receive certain relationship benefits and recognition. But in the process of making and implementing these claims, other norms come into play. In some cases these are norms – such as commitment and responsibility – explicitly cited and drawn on by law reform campaigners to give their demands greater authority and legitimacy. In other cases, norms anchoring familial and relationship practices work to inflect and refract the pursuit of same-sex spousal status in less helpful ways. In exploring this process, my focus is the political effects of this normative incursion. Two normative principles that proved central to the refraction of spousal recognition reforms are proper place and the public/private. In their prevailing discursive and citational structure, with their stress on propriety, propinquity and akinship, both work to limit the normative impact of the reforms. They also, I argue, inflect and narrow the egalitarian impulse of extending spousal status by channelling benefits to those whose lives, for reasons of class, race or life choices, already dovetail most comfortably with dominant readings of the proper and the public/private.

The fortification of dominant normative readings not only occurs through law reform campaigns. Another common driver is the articulation of harm. Harm is a core structure for determining which forms of diversity and inequality are acceptable, as ongoing struggles over the right of religious organisations to exclude gay personnel demonstrate. Harm is largely seen as a self-evident truth, yet closer examination demonstrates the ways in which determining the character and recipient of harm depends on, and in turn entrenches, prevailing values. This is apparent in the way in which sex forced by a husband on his wife was not legally defined as a harm throughout most of the twentieth century in Britain, while rape has in many historical and national contexts been read as an (economic) injury to a woman's father, husband or owner rather than to the woman herself. In exploring the ways in which claims of harm sustain the norms that anchor them, my discussion, in chapter 6, focuses on nuisance. Nuisance utterances, policy and law have proved to be particularly effective in naturalising, and thereby reinforcing, particular normative readings. In Britain, nuisance discourse has pinned freedom to self-interest, responsibility and proper place in ways that draw from, and support, prevailing social dynamics and inequalities.

At the same time normative principles are not unchanging. What proper place or the public/private means can be reinflected through creating new citational or referential loops. Such loops can work by recentring normative principles such as equality or diversity. As Mouffe (2000: 10) argues in her discussion of the relationship between democracy and liberalism, 'once the articulation of the two principles has been effectuated – even if in a precarious way – each of them changes the identity of the other'. In the case of spousal law reform, how changes are argued for, operationalised and inhabited can make a difference to the norms they refract. For while these norms provide a lens through which reforms are rendered intelligible, the lens can also be adjusted. Nuisance, similarly, illustrates the capacity of norms to be rearticulated. Reversal of one norm, such as responsibility, can offer leverage to 'flipping over' others, for instance, through the strategic identification of police officers who interfere, or people who pass by without giving money, as nuisances or anti-social to the person begging. By repositioning responsibility as something owed to strangers without, normative principles of freedom become articulated to assistance rather than the right to be left alone.

Nuisance also challenges the ways in which dominant normative principles circulate in a different respect. Aside from strategically articulating normative principles in consciously oppositional ways, social constituencies can come tactically to occupy the positions dominant nuisance discourse creates. Historically this has taken the form of obstructing or disrupting economic and legal processes, transmitting displeasing sights, sounds and smells, or has entailed an explicit politics of inappropriate placing – transgressing roles and boundaries. To the extent that transgressive conduct is read as a nuisance, it may have little immediate strategic effect on the way in which norms and social inequalities are understood. Indeed, it may precipitate a clamour of fortificatory activity as existing norms are resecured. However, over the longer term, normative change may emerge, particularly in relation to norms of appropriateness, as what were once transgressions of racial, gender and sexual boundaries no longer divert attention.

In my discussion of nuisance conduct, I suggested that transgressions frequently arose in the course of creating intentional communities, and it is here that innovative reworkings of normative principles have particularly occurred. In chapter 8, I explore the establishment of prefigurative communities; my analysis there focuses on the relationship between the new social pathways developed, their environment and the boundaries that secure them. My starting point in this discussion was the difficulty of sustaining counter-normative ways of being in the wider environment when such pathways were perceived as foolish, irrational or naive, or rendered exhausting and costly. This does not mean, however, that such pathways fail to occur. Counter-normative pathways are continually being established on an individual and socially networked basis and many may develop and flourish – at least for a while. However, without a conducive environment or a strong and steady habitus, the thin support that comes from superordinate orders, compatible norms or conducive resource allocations, for example (but not from all three together), means that such practices are likely to prove hard to sustain over time.

Creating bounded communities, by contrast, enables social environments to be established in which counter-normative conduct makes sense. However, this process is invariably partial, structured by processes of permeation, convergence and isomorphism. Writing on progressive or intentional communities has adopted an equivocal perspective on the effects of these processes for communities. In chapter 8 I outline some of the problems permeability causes. At the same time, the capacity of people in particular to enter and leave a community is often essential to the impression such communities can have. This impression on the mainstream comes from showcasing alternative ways of living to visitors, providing a concentration of utterances that are transmitted beyond the community's walls, and facilitating political mobilisation through the movement of people, both within the community and beyond (Rigby 1974: 33). These effects are important. However, in closing, I want to consider two distinctive features and challenges offered by the kinds of communities discussed. The first concerns the experience of articulating organising principles in counter-normative ways; the second addresses communities' wider impact.

Communities such as Greenham, LETS and Summerhill not only demonstrate the ability of counter-normative principles to become embedded, they also demonstrate how their meaning is generated by, and made intelligible through, alternative citational processes. The ways in which equality, for instance, is rendered meaningful depends on its relationship to other norms – both those that give it texture as well as those against which it is distinguished. Although the process of rearticulating norms is carried out widely within progressive movements, communities play a distinctive role because the disjuncture between imagined and inhabited pathways is compressed. In other words, such rearticulations do not function as rhetorical ideals, they are also embedded in the institutions and social dynamics of a community and its pathways.

This is not a closed and seamless process. Normative principles can rub uneasily against each other: this is apparent in LETS, where norms of publicity, oriented to equality and regard for strangers, buffeted against norms of localism and individual autonomy. One effect such friction produces is greater diversity, as people pursue pathways anchored in different normative citational loops. This is the dilemmatic environment raised earlier, where similarly positioned actors face an ambivalent social logic. This ambivalence is not simply a product of competing norms. Resource allocations, institutional practices and social demands also shape the nature of the dilemma confronted. So LETS members may struggle to reach a satisfactory solution to the question of availability, price, quality and responsiveness, as they seek to integrate – the not always compatible – norms of equality, commitment, fairness, and trust, with personal concerns about safety, economic welfare, social wellbeing and fatigue, and with other non-LETS-based demands and claims upon them. In these kinds of dilemmatic situations, whichever compromise is reached, actors feel the tug of the alternative. While this can produce greater diversity, as the LETS context illustrates, and lead to the sustainability of minority practices within the niches enabled by dilemmatic environments, it also risks precipitating an exodus of members as they find the inevitable pull of alternatives unduly wearing and unsatisfactory.

Another effect of the dilemmatic arises when the pathways crafted within a counter-normative environment fail to match those actually inhabited. This point has been made with some force by a number of scholars including Jo Freeman (1984), whose analysis of the informal power structures within ostensibly 'structureless' feminist communities proved particularly influential. Others have explored the way in which dysfunctional or non-normative practices can become routinised. LETS members I interviewed described the problems caused by members' carelessness in relation to appointments and commitments; several people identified a tendency for members to treat LETS agreements less seriously than trades agreed in national currency. In the case of Greenham, non-normative pathways resembled those identified by other commune scholars, namely members failing to 'do their share' and treating collective goods with disregard. Non-normative practices are often a site of strain for communities. Rarely presented to visitors or publicly 'showcased', such pathways can cause a community eventually to implode. More commonly, it precipitates a return to practices, such as individual ownership and differentiated incomes, that replicate systems and norms of the wider society (e.g., Spiro 1963).

One community which seems from my research to have adopted a different approach to non-normative pathways is that of Summerhill School. Their establishment of complex lawmaking and dispute resolution structures, anchored in a recognition of conflict and antisocial behaviour, allows the school to demonstrate publicly and confidently their ability to deal with nuisance or destructive conduct. This contrasts with LETS, in particular, where, with few mediation procedures, scheme members seem very reluctant to discuss or acknowledge the existence of non-normative behaviour. Managing the challenge posed by both conflict and 'behaving badly' is crucial for alternative communities' survival. At Summerhill the legitimacy of diverging opinions, for instance, is written into the core of its decision-making processes, while breaking laws is subjected to a non-punitive system of restitution and deterrence that seeks to preserve as far as possible the dignity and sense of belonging of individual members (Neill 1968). In other communities conflict is less equably recognised or resolved. As a consequence, while Summerhill is able quietly to contain non-normative conduct so that it does not threaten core school values, in other communities such conduct has generated far more passion and heat.

In saying this, I do not want to suggest that isomorphic legal structures are necessary elements in alternative communities. Rather, what Summerhill identifies is the value of being able to respond to dissent and antagonism – or agonism (see Mouffe 2000). It also suggests the value of not lumping together all conduct seen as non-normative. For within this category, as I discussed earlier, there will be manageable non-normative practices, pathways that jeopardise the community and productive challenges or alternatives to community norms. How these are distinguished may depend on whether non-normative conduct is adopted by a minority or becomes more pervasive, and whether it is reproduced or not. Yet none of these are hard and fast divisions. More generally, the pursuit of non-normative or oppositional pathways is a space of enormous potential as well as concern. Neglected by many commentators and community structures alike, it is as important to the political possibilities radical communities engender as the hegemonic practices they more publicly and willingly constitute and embrace.

The second point I wish to highlight concerns the impact of community practices on wider institutional structures and norms. This impact flags up the role of the margins, and of boundary actors within communities, who perform the double manoeuvre described in chapter 8 of internal and external critique, while pursuing a new hegemony within and without. Described in this way, it may sound as if a community's impact depends on the consciousness and strategy of the actors that comprise it. But communities can trouble dominant practices and create the impression of new ones through the exodus and flow of incompatible cultural and economic capital and norms, to some extent regardless of members' intentions. These may not receive a welcome response, but their impact also does not depend on their doing so. Rather, I want to argue, a major reason why alternative practices leave an impression comes from their iterability.

Conventional forms of political protest deploy a repertoire of strategic action that involves marches, demonstrations, strikes, letter-writing, meetings and publicity. Although these actions may involve repetition, particularly where protests seek to be obstructive, this is usually temporary. The emphasis on generating surprise is heightened by the priority placed on obtaining media attention which responds to novel, enterprising and 'one-off' actions. In contrast, pathways established and maintained by community environments, whether in the form of LETS trading, Summerhill's non-compulsory classes, or Greenham protestors' disregard for the law, impress on institutional forces through their repetition. Such pathways may remain too consciously oppositional or counter-normative to become routinised, but they demand a response because they cause continual friction and do not immediately disappear. However, it is not enough that communities engage in repetitive and routinised behaviour. It is largely because their boundaries are permeable that they are able to indent on wider life. Isomorphic practices, such as Summerhill's laws and LETS' accounting systems, also help, since they render community norms intelligible without the need for thick translation processes.

But if permeability is important to and helps to feed a community's wider impact, what counter-strategies does this, in turn, generate? In particular, do wider institutional and governmental forces seek to limit a community's effects through sealing it up? This question reveals the double-sided quality of permeability. While I have argued that communities *need* boundaries – normative and symbolic as well as spatial – to secure a non-hegemonic social environment, boundaries may also be 'stopped up' against community attempts to increase their porosity. The struggle this generates can be seen in relation to Greenham peace camp. At Greenham, the use of the law and the custodial system to 'close down' the camp generated an outflow of publicity and actors that gave oxygen to Greenham's critique of militarism and wider gender norms. By contrast, it was the stopping-up of informational flows that sabotaged Greenham's impact as the mass media lost interest in the camp in the late 1980s. Summerhill reveals a similar process. While the government failed, more than a decade post-Greenham, in its use of the law to close the school down, its lack of interest in the school and general refusal to recognise Summerhill's educational contribution resecured the normative boundary around it – a boundary that had threatened to become more porous as a result of the media and educational attention generated by the legal conflict.

The capacity of governmental and institutional forces to blockade community practices raises the question as to what can be done by communities and others to resist this process. Communities often feel so beleaguered that they turn inwards. Wider progressive and radical forces forfeit any impact they might have by leaving communities to their own devices. Left rhetoric and discourses that see communities as marginal at best, but more usually as diversionary and self-indulgent, facilitate hostile forces in their efforts to seal communities up. Greenham peace camp, to some extent, broke through this isolation by mobilising and exciting sympathetic forces at a distance. Summerhill has also generated a wider interest and commitment that has waxed and waned over its eighty-year history. But despite these instances, in the main outsiders to prefigurative communities have played only a minimal role in resisting the border efforts of official forces. Some might argue that this is the result of a community-building logic: that those beyond feel alienated, with little interest in the success of the political project mobilised by those within. Perhaps this is true. But since communities play a key part in the creation of counter-normative practices – practices which not only posit alternatives but also provide a practical critique of the mainstream – how to strengthen and amplify their capacity to dig into mainstream institutional and everyday life remains a key, underrated challenge.

Bibliography

Abrams, P. and McCulloch, A., with Abrams, S. and Gore, P., 1976, *Communes, Sociology and Society*, Cambridge: Cambridge University Press.

Acker, J., 2000, 'Revisiting class: thinking from gender, race, and organizations', *Social Politics* 7, 192–214.

Adams, T., 1998, 'Combining gender, class and race: structuring relations in the Ontario dental profession', *Gender and Society* 12, 578–97.

Adams, T., 2000, *A Dentist and a Gentleman: Gender and the Rise of Dentistry in Ontario*, Toronto: University of Toronto Press.

Adkins, L. 1994, *Gendered Work: Sexuality, Family and the Labour Market*, Buckingham: Open University Press.

Adler, S. and Brenner, J., 1992, 'Gender and space: lesbians and gay men in the city', *Journal of Urban and Regional Research* 16, 24–34.

Alarcón, N., 1990, 'The theoretical subject(s) of *This Bridge Called My Back* and Anglo-American feminism', in G. Anzaldúa (ed.), *Making Face, Making Soul/Haciendo Caras*, San Francisco: Aunt Lute Foundation.

Aldrich, H. and Mindlin, S., 1978, 'Uncertainty and dependence: two perspectives on environment', in L. Karpik (ed.), *Organization and Environment: Theory, Issues and Reality*, London: Sage.

Aldridge, T. et al., 2001, 'Recasting work: the example of local exchange trading schemes', *Environment and Planning A* 31, 2033–51.

Amos, V. and Parmar, P., 1984. 'Challenging imperial feminism', *Feminist Review* 17, 3–19.

Anzaldúa, G., 1983, 'La preieta', in C. Moraga and G. Anzaldúa (eds.), *This Bridge Called My Back*, New York: Kitchen Table: Women of Color Press.

Anzaldúa G. (ed.), 1990, *Making Face, Making Soul/Haciendo Caras*, San Francisco: Aunt Lute Foundation.

Appiah, K., 1994, 'Identity, authenticity, survival: multicultural societies and social reproduction', in A. Gutmann (ed.), *Multiculturalism: Examining the Politics of Recognition*, Princeton: Princeton University Press.

Appleton, M., 2000, *A Free Range Childhood: Self Regulation at Summerhill School*, Brandon, VT: The Foundation for Educational Renewal, Inc.

Avido, D., 1986, 'Health issues relating to "passive" smoking', in R. Tollison (ed.), *Smoking and Society: Toward a More Balanced Assessment*, Lexington, Mass: D.C. Heath & Co.

Azoulay, K., 1997, *Black, Jewish and Interracial: It's Not the Color of Your Skin but the Race of Your Kin, and Other Myths of Identity*, Durham, NC: Duke University Press.

Ball, S., 1990, *Markets, Morality and Equality in Education*, Hillcole Group Paper 5, London: Tufnell Press.

Ball, S., 1993, 'Education policy, power relations and teachers' work', *British Journal of Educational Studies* 41, 106–21.

Ball, W. and Solomos, J. (eds.), 1990, *Race and Local Politics*, Basingstoke: Macmillan.

Barley, S. and Tolbert, P., 1997, 'Institutionalization and structuration: studying the links between action and institution', *Organization Studies* 18, 93–117.

Barrett, M. and McIntosh, M., 1980, 'The "family wage": some problems for socialists and feminists', *Capital and Class* 11, 51–72.

Barry, J. and Proops, J., 2000, *Citizenship, Sustainability and Environmental Research*, Cheltenham: Edward Elgar.

Bauman, Z., 1993, *Postmodern Ethics*, Oxford: Blackwell.

Bauman, Z., 1996, 'On communitarians and human freedom or, how to square the circle', *Theory, Culture and Society* 13, 79–90.

Beckett, F., 2000, 'The loaded dice – private sector involvement in education', *Education Review* 13, 49–53.

Bell, D. and Binnie, J., 2000, *The Sexual Citizen: Queer Politics and Beyond*, Cambridge: Polity.

Bell, L., 1999, 'Back to the future: the development of educational policy in England', *Journal of Educational Administration* 37, 200–28.

Ben-Tovim, G. et al., 1986, *The Local Politics of Race*, Basingstoke: Macmillan.

Benton, L., 1994, 'Beyond legal pluralism: towards a new approach to law in the informal sector', *Social and Legal Studies* 3, 223–42.

Berger, P., 1986, 'A sociological view of the antismoking phenomenon', in R. Tollison (ed.), *Smoking and Society: Toward a More Balanced Assessment*, Lexington, MA: D.C. Heath & Co.

Berlin, I., 1991, 'Two concepts of liberty', in D. Miller (ed.), *Liberty*, Oxford: Oxford University Press.

Bey, H., 1991, *T.A.Z. The Temporary Autonomous Zone, Ontological Anarchy, Poetic Terrorism*, Brooklyn, NY: Autonomedia.

Bickford, S., 1997, 'Anti-anti-identity politics: feminism, democracy, and the complexities of citizenship', *Hypatia* 12, 111–31.

Bickley, J. and Beech, A., 2002, 'An investigation of the Ward and Hudson pathways model of the sexual offense process with child abusers', *Journal of Interpersonal Violence* 17, 371–93.

Binnie, J. and Bell, D., 2003, 'Rethinking sexual citizenship in the city', presented at the Association of American Geographers, Annual Meeting, New Orleans, 5–8 March 2003.

Blackwood, C., 1984, *On the Perimeter*, London: Flamingo.

Boddy, M. and Fudge, C. (eds), 1984, *Local Socialism?* London: Macmillan.

Boga, T., 1994, 'Turf wars: street gangs, local governments, and the battle for public space', *Harvard Civil Rights–Civil Liberties Law Review* 29, 477–503.

Bohman, J., 1999, 'Citizenship and norms of publicity', *Political Theory* 27, 176–202.

Bone, R., 1986, 'Normative theory and legal doctrine in American nuisance law: 1850 to 1920', *Southern California Law Review* 59, 1101–226.

Bornstein, K., 1994, *Gender Outlaw: On Men, Women and the Rest of Us*, London: Routledge.

Bourdieu, P., 1977, *Outline of a Theory of Practice*, Cambridge: Cambridge University Press.

Bourdieu, P., 1984, *Distinction: A Social Critique of the Judgement of Taste*, London: Routledge and Kegan Paul.

Bourdieu, P., 1992, *The Logic of Practice*, Cambridge: Polity.

Bourdieu, P., 1993, *Sociology in Question*, London: Sage.

Bourdieu, P., 1994, *In Other Words*, Cambridge: Polity.

Bourdieu, P. and Wacquant, L., 1992, *An Invitation to Reflexive Sociology*, Chicago: University of Chicago Press.

Bouthillette, A.-M., 1997, 'Queer and gendered housing: a tale of two neighbourhoods in Vancouver', in G. Ingram et al. (eds), *Queers in Space*, Seattle: Bay Press.

Bowring, F., 1998, 'LETS: an eco-socialist initiative', *New Left Review* 232, 91–111.

Boyd, S., 1996, 'Best friends or spouses? Privatisation and the recognition of lesbian relationships in *M.* v. *H.*', *Canadian Journal of Family Law* 13, 321–41.

Boyd, S., 1999, 'Family, law and sexuality: feminist engagements', *Social and Legal Studies* 8, 369–90.

Boyd, S. and Young, C., 2003, '"From same-sex to no sex"? Trends towards recognition of (same-sex) relationships in Canada', *Seattle Journal for Social Justice* 1, 757–93.

Brenner, J., 1974, 'Nuisance law and the industrial revolution', *Journal of Legal Studies* 3, 403–33.

Breslow, L., 1982, 'Control of cigarette smoking from a public policy perspective', *Annual Review of Public Health* 3, 129–51.

Brewer, R., 1993, 'Theorizing race, class and gender', in S. James and A. Busia (eds.), *Theorizing Black Feminisms: The Visionary Pragmatism of Black Women*, London: Routledge.

Brickell, C., 2001, 'Whose "special treatment"? Heterosexism and the problems with liberalism', *Sexualities* 4, 211–35.

Bright, S., 2001, 'Liability for the bad behaviour of others', *Oxford Journal of Legal Studies* 21, 311–30.

Brown, M., 1995, 'Sex, scale and the "new urban politics": HIV-prevention strategies from Yaletown, Vancouver', in D. Bell and G. Valentine (eds.), *Mapping Desire*, London: Routledge.

Brown, M., 1997, *RePlacing Citizenship: AIDS Activism and Radical Democracy*, New York: Guilford Press.

Brown, M., 2000, *Closet Space: Geographies of Metaphor from the Body to the Globe*, London: Routledge.

Brownworth, V., 1996, 'Tying the knot or the hangman's noose', *Journal of Gay, Lesbian and Bisexual Identity* 1, 91–8.

Brunsson, N. and Olsen, J., 1993, *The Reforming Organization*, London: Routledge.

Butler, J., 1990, *Gender Trouble: Feminism and the Subversion of Identity*, New York: Routledge.

Butler, J., 1993, *Bodies that Matter*, New York: Routledge.

Butler, J., 1997a, *Excitable Speech: A Politics of the Performative*, New York: Routledge.

Butler, J., 1997b, 'Critically queer', in Shane Phelan (ed.), *Playing with Fire: Queer Politics, Queer Theories*, New York: Routledge.

Butler, J., 1998, 'Merely cultural', *New Left Review* 227, 33–44.

Butler, S. and Weatherley, R., 1995, 'Pathways to homelessness among middle-aged women', *Women and Politics* 15, 1–22.

Calhoun, C., 1993, 'Habitus, field, and capital: the question of historical specificity', in C. Calhoun et al. (eds.), *Bourdieu: Critical Perspectives*, Cambridge: Polity.

Calhoun, C., 1999, 'Nationalism, political community and the representation of society or, why feeling at home is not a substitute for public space', *European Journal of Social Theory* 2, 217–31.

Califia, P., 1997, *Sex Changes: The Politics of Transgenderism*, San Francisco: Cleis Press.

Campbell, C., 1996, 'Detraditionalization, character and the limits to agency', in P. Hellas et al. (eds.), *Detraditionalization*, Oxford: Blackwell.

Campbell, S., 1995, 'Gypsies: the criminalisation of a way of life?', *Criminal Law Review* 28–37.

Carabine, J., 1995, "Invisible sexualities: sexuality, politics and influencing policy-making", in A. Wilson (ed.), *A Simple Matter of Justice?* London: Cassell.

Card, C., 1996, 'Against marriage and motherhood', *Hypatia* 11, 1–23.

Carrington, C., 1999, *No Place Like Home: Relationships and Family Life among Lesbians and Gay Men*, Chicago: University of Chicago Press.

Carroll, W. and Ratner, R., 2001, 'Sustaining oppositional cultures in "post-socialist" times: a comparative study of three social movement organisations', *Sociology* 39, 605–29.

Chambers, D., 1996, 'What if? The legal consequences of marriage and the legal needs of lesbian and gay male couples', *Michigan Law Review* 95, 447–91.

Cheyette, B., 1995, *Constructions of 'the Jew' in English Literature and Society: Racial Representations, 1875–1945*, Cambridge: Cambridge University Press.

Clarke, J. and Newman, J., 1997, *The Managerial State*, London: Sage.

Cohen, I., 1988, 'Structuration theory and social praxis', in A. Giddens and J. Turner (eds.), *Social Theory Today*, Cambridge: Polity.

Comacchio, C., 1999, *The Infinite Bonds of Family: Domesticity in Canada, 1850–1940*, Toronto: University of Toronto Press.

Combahee River Collective, 1979, 'A black feminist statement', in Z. Eisenstein (ed.), *Capitalist Patriarchy and the Case for Socialist Feminism*, New York: Monthly Review Press.

Conaghan, J., 1993, 'Harassment and the law of torts: *Khorasanjian* v. *Bush*', *Feminist Legal Studies* 13, 189–97.

Conaghan, J., 1996, 'Gendered harms and the law of tort: remedying (sexual) harassment', *Oxford Journal of Legal Studies* 16, 407–37.

Conaghan, J. and Mansell, W., 1999, *The Wrongs of Tort*, London: Pluto Press (2nd edn).

Connolly, W., 1995, *The Ethos of Pluralization*, Minneapolis: University of Minnesota Press.

Cooper, D., 1994, *Sexing the City: Lesbian and Gay Politics within the Activist State*, London: Rivers Oram.

Cooper, D., 1995a, *Power in Struggle: Feminism, Sexuality and the State*, Buckingham: Open University Press.

Cooper, D., 1995b, 'Defiance and non-compliance: religious education and the implementation problem', *Current Legal Problems* 48, 253–79.

Cooper, D., 1995c, 'Local government legal consciousness in the shadow of juridification', *Journal of Law and Society* 22, 506–26.

Cooper, D., 1998a, *Governing Out of Order: Space, Law and the Politics of Belonging*, London: Rivers Oram.

Cooper, D., 1998b, 'Regard between strangers: diversity, equality and the re-construction of public space', *Critical Social Policy* 18, 465–92.

Cooper, D., 2001, 'Boundary harms: from community protection to a politics of value: the case of the Jewish eruv', in G. Hughes et al. (eds.), *Community Safety, Crime Prevention and Social Control*, London: Sage.

Cooper, D., 2002, 'Imagining the place of the state: where governance and social power meet', in D. Richardson and S. Seidman (eds.), *Handbook of Lesbian and Gay Studies*, London: Sage.

Cooper, D. and Monro, S., 2003, 'Governing from the margins: queering the state of local government?' *Contemporary Politics* 9: 229–55.

Corbin, A., 1994, *The Foul and the Fragrant: Odor and the French Social Imagination*, London: Picador.

Cornell, D., 1998, *At the Heart of Freedom*, Princeton, NJ: Princeton University Press.

Cossman, B. 1997, 'Feminist fashion or morality in drag? The sexual subtext of the *Butler* decision', in B. Cossman et al. (eds.), *Bad Attitude/s on Trial: Pornography, Feminism, and the Butler Decision*, Toronto: Toronto University Press.

Cossman, B. et al. (eds.), 1997, *Bad Attitude/s on Trial: Pornography, Feminism, and the Butler Decision*, Toronto: Toronto University Press.

Cox, B., 1997a, 'The lesbian wife: same-sex marriage as an expression of radical and plural democracy', *California Western Law Review* 33, 155–67.

Cox, B., 1997b, 'A (personal) essay on same-sex marriage', in R. Baird and S. Rosenbaum (eds.), *Same-Sex Marriage. The Moral and Legal Debate*, Amherst, NY: Prometheus Books.

Crane, B., 2000, 'Filth, garbage, and rubbish: refuse disposal, sanitary reform, and nineteenth-century yard deposits in Washington, D.C.', *Historical Archaeology* 34, 20–38.

Crenshaw, K., 1989, 'Demarginalizing the intersection of race and sex: a black feminist critique of antidiscrimination doctrine, feminist theory and antiracist politics', *University of Chicago Legal Forum* 139–167.

Cresswell, T., 1996, *In Place/Out of Place: Geography, Ideology, and Transgression*, Minneapolis: University of Minnesota Press.

Croall, J., 1997, *LETS Act Locally*, London: Calouste Gulbenkian Foundation.

Cross, G., 1995, 'Does only the careless polluter pay? A fresh examination of the nature of private nuisance', *Law Quarterly Review* 111, 445–74.

Currah, P., 1997, 'Politics, practices, publics: identity and queer rights', in S. Phelan (ed.), *Playing with Fire: Queer Politics, Queer Theories*, New York: Routledge.

Darling, J., 1992, 'A. S. Neill on democratic authority: a lesson from Summerhill?', *Oxford Review of Education* 18, 45–57.

Davis, A., 1981, *Women, Race and Class*, New York: Random House.

Davis, T., 1995, 'The diversity of gay politics and the redefinition of sexual identity and community in urban spaces', in D. Bell and G. Valentine (eds.), *Mapping Desire*, London: Routledge.

de Certeau, M., 1984, *The Practice of Everyday Life*, Berkeley: University of California Press.

de Lauretis, T., 1990, 'Eccentric subjects: feminist theory and historical consciousness', *Feminist Studies* 16, 115–50.

Dean, C., 1994, 'Gay marriage: a civil right', *Journal of Homosexuality* 27, 111–15.

Deem, R. et al., 1994, 'Governors, schools and the miasma of the market', *British Educational Research Journal* 20, 535–49.

Demaine, J., 1988, 'Teachers' work, curriculum and the new right', *British Journal of Sociology of Education* 9, 247–63.

DiMaggio, P. and Powell, W., 1991, 'Introduction', in W. Powell and P. DiMaggio (eds.), *The New Institutionalism in Organizational Analysis*, Chicago: University of Chicago Press.

Dobson, R., 1993, *Bringing the Economy Home from the Market*, London: Black Rose Books.

Doherty, B., 1996, 'Paving the way: the rise of direct action against roadbuilding and the changing character of British environmentalism', *Alternative Futures and Popular Protest II* (conference papers, vol. I), Manchester: Manchester Metropolitan University.

Dreyfus, H. and Rabinow, P., 1983, *Michel Foucault: Beyond Structuralism and Hermeneutics*, Chicago: University of Chicago Press (2nd edn).

Dumm, T., 1996, *Michel Foucault and the Politics of Freedom* London: Sage.

Dunn, J. and Hayes, M., 1999, 'Identifying social pathways for health inequalities. The role of housing', *Annals of the New York Academy of Sciences* 896, 399–402.

Dworkin, R., 1981, 'What is equality? Part 1: equality of welfare', *Philosophy and Public Affairs* 10, 185–246.

Dworkin, R., 1985, *A Matter of Principle*, Cambridge, MA: Harvard University Press.

Dworkin, R., 1986, *Law's Empire*, Cambridge, MA: Harvard University Press.

Dworkin, R., 1994, *Taking Rights Seriously*, London: Duckworth (7th edn).

Dworkin, R., 1995, 'Foundations of liberal equality', in S. Darwall (ed.), *Equal Freedom (Selected Tanner Lectures on Human Values)*, Ann Arbor, MI: University of Michigan Press.

Dworkin, R., 2000, *Sovereign Virtue: The Theory and Practice of Equality*, Cambridge, MA: Harvard University Press.

Eaton, M., 1995, 'Homosexual unmodified: speculations on law's discourse, race, and the construction of sexual identity', in D. Herman and C. Stychin (eds.), *Legal Inversions: Lesbians, Gay Men, and the Politics of Law*, Philadelphia: Temple University Press.

Edensor, T., 1998, *Tourists and the Raj: Performance and Meaning at a Symbolic Site*, London: Routledge.

Edensor, T., 1999, 'Moving through the city', in D. Bell and A. Haddour (eds.), *City Visions*, Longman: London.

Eichler, M., 1997, *Family Shifts: Families, Policies, and Gender Equality*, Oxford: Oxford University Press.

Eisenberg, A., 2003, 'Diversity and equality: three approaches to cultural and sexual difference', *The Journal of Political Philosophy* 11, 41–64.

Eisenstein, Z., 1979, 'Developing a theory of capitalist patriarchy and socialist feminism', in Z. Eisenstein (ed.), *Capitalist Patriarchy and the Case for Socialist Feminism*, New York: Monthly Review Press.

Eisenstein, Z., 1994, *The Color of Gender: Reimaging Democracy*, Berkeley: University of California Press.

Eisenstein, Z., 1996 'Equalizing privacy and specifying equality', in Nancy J. Hirschmann and Christine Di Stefano (eds.), *Revising the Political: Feminist Reconstructions of Traditional Concepts in Western Theory*, Boulder, CO: Westview Press.

Ellickson, R., 1973, 'Alternatives to zoning: covenants, nuisance rules, and fines as land use controls', *University of Chicago Law Review* 40, 681–781.

Ellickson, R., 2001, 'Controlling chronic misconduct in city spaces: of panhandlers, skid rows, and public-space zoning', in N. Blomley et al. (eds.), *The Legal Geographies Reader*, Oxford: Blackwell.

Ellis, K. et al., 1999, 'Needs assessment, street-level bureaucracy and the new community care', *Social Policy and Administration*, 33, 262–80.

Emberley, J. and Landry, D., 1989, 'Coverage of Greenham and Greenham as "coverage"', *Feminist Studies* 15, 485–98.

Epstein, D., 1996, 'What's in a ban? The popular media, "Romeo and Juliet" and compulsory heterosexuality', in D. Steinberg et al. (eds.), *Border Patrols: Policing Sexual Boundaries*, London: Cassell.

Eskridge, W., 2001, *Equality Practice: Civil Unions and the Future of Gay Rights*, New York: Routledge.

Ettelbrick, P., 1997, 'Since when is marriage a path to liberation?' in R. Baird and S. Rosenbaum (eds.), *Same-Sex Marriage. The Moral and Legal Debate*, Amherst, NY: Prometheus Books.

Ewick, P. and Silbey, S., 1992, 'Conformity, contestation, and resistance: an account of legal consciousness', *New England Law Review* 26, 731–49.

Fawcett, B., 1996, 'Postmodernism, feminism and disability', *Scandinavia Journal of Social Welfare* 5, 259–67.

Feinberg, J., 1985, *The Moral Limits of the Criminal Law, Vol. 2, Offense to Others*, New York: Oxford University Press.

Feinhandler, S., 1986, 'The social role of smoking', in R. Tollison (ed.), *Smoking and Society: Toward a More Balanced Assessment*, Lexington, MA: D. C. Heath & Co.

Feldman, M., 2000, 'Organizational routines as a source of continuous change', *Organization Science* 11, 611–29.

Fenstermaker, S. and West, C. (eds.), 2002, *Doing Gender, Doing Difference: Inequality, Power, and Institutional Change*, New York: Routledge.

Ferguson, A., 1991, *Sexual Democracy: Women, Oppression, and Revolution*, Boulder, CO: Westview Press.

Fergusson, R., 1994, 'Managerialism in education', in J. Clarke et al. (eds.), *Managing Social Policy*, London: Sage.

Finch, S. et al., 1986, 'Socialist-feminisms and Greenham', *Feminist Review* 23, 93–100.

Fishman, A., 1995, 'Modern orthodox Judaism: a study in ambivalence', *Social Compass* 42, 89–95.

Flax, J., 1987, 'Postmodernism and gender relations in feminist theory', *Signs* 12, 621–43.

Flax, J., 1992, 'Beyond equality: gender, justice and difference', in G. Block and S. James (eds.), *Beyond Equality and Difference*, London: Routledge.

Flax, J., 1995, 'Race/gender and the ethics of difference: a reply to Okin's "gender inequality and cultural differences"', *Political Theory* 23, 500–10.

Florida, R., 2002, *The Rise of the Creative Class*, New York: Basic Books.

Foucault, M., 1980, *Power/Knowledge*, New York: Pantheon.

Foucault, M., 1981, *The History of Sexuality*, London: Pelican.

Foucault, M., 1983, 'The subject and power', in H. Dreyfus and P. Rabinow (eds.), *Michel Foucault: Beyond Structuralism and Hermeneutics*, Chicago: University of Chicago Press (2nd edn).

Fraser, N., 1989, *Unruly Discourses: Power, Discourse, and Gender in Contemporary Social Theory*, Cambridge: Polity Press.

Fraser, N., 1997, *Justice Interruptus: Critical Reflections on the 'Postsocialist' Condition*, New York: Routledge.

Fraser, N., 1998, 'Social justice in the age of identity politics: redistribution, recognition, and participation', in G. Peterson (ed.), *The Tanner Lectures on Human Values*, 19: 1–67, Utah: University of Utah Press.

Fraser, N., 2001, 'Recognition without ethics?' *Theory, Culture and Society*, 18, 21–42.

Freeman, J., 1984, 'The tyranny of structurelessness', in *Untying the Knot: Feminism, Anarchism and Organisation*, London: Dark Star Press and Rebel Press.

Freund, P., 2001, 'Bodies, disability and spaces: the social model and disabling spatial organisations', *Disability and Society* 16, 689–706.

Fridman, G., 1954, 'Motive in the English law of nuisance', *Virginia Law Review* 40, 583–95.

Galeotti, A., 1993, 'Citizenship and equality: the place for toleration', *Political Theory* 21, 585–605.

Galeotti, A., 2002, *Toleration as Recognition*, Cambridge: Cambridge University Press.

Garner, D., 1977, 'Cigarettes and welfare reform', *Emory Law Journal* 26, 269–335.

George, R., 1998a, 'Recycling: long routes to and from domestic fixes', in R. George (ed.), *Burning Down the House: Recycling Domesticity*, Boulder, CO: Westview Press.

George, R., 1998b, 'Homes in the empire, empires in the home', in R. George (ed.), *Burning Down the House: Recycling Domesticity*, Boulder, CO: Westview Press.

Gewirtz, S. et al., 1993, 'Values and ethics in the education market place: the case of Northwark Park', *International Studies in Sociology of Education* 3, 233–54.

Glenn, E. N., 1985, 'Racial ethnic women's labor: the intersection of race, gender and class oppression', *Review of Radical Political Economics* 17, 86–108.

Godard, J., 2002, 'Institutional environments, employer practices, and states in liberal market economies', *Industrial Relations* 41, 249–85.

Goldberg-Hiller, J., 2002, *The Limits to Union: Same Sex Marriage and the Politics of Civil Rights*, Ann Arbor: University of Michigan Press.

Goodin, R., 1989, *No Smoking: The Ethical Issues*, Chicago: University of Chicago Press.

Gotell, L., 1997, 'Shaping *Butler*: The new politics of anti-pornography', in B. Cossman et al. (eds.), *Bad Attitude/s on Trial*, Toronto: University of Toronto Press.

Gottfried, H., 1998, 'Beyond patriarchy? Theorising gender and class', *Sociology* 32, 451–68.

Green, V., 1996, 'Same-sex adoption: an alternative approach to gay marriage in New York', *Brooklyn Law Journal* 62, 399–429.

Greenwalt, K., 1995, *Fighting Words: Individuals, Communities, and Liberties of Speech*, Princeton: Princeton University Press.

Greenwood, R. and Hinings, C., 1996, 'Understanding radical organizational change: bringing together the old and the new institutionalism', *Academy of Management Review* 21, 1022–54.

Grendstad, G. and Selle, P., 1995, 'Cultural theory and the new institutionalism', *Journal of Theoretical Politics* 7, 5–27.

Gribble, D., 1998, *Real Education: Varieties of Freedom*, Bristol: Libertarian Education.

Grube, J., 1997, '"No more shit": the struggle for democratic gay space in Toronto', in G. Ingram et al. (eds.), *Queers in Space*, Seattle: Bay Press.

Gyford, J., 1985, *The Politics of Local Socialism*, London, George Allen & Unwin.

Hage, J., 1978, 'Toward a synthesis of the dialectic between historical-specific and sociological-general models of the environment', in L. Karpik (ed.), *Organization and Environment: Theory, Issues and Reality*, London: Sage.

Halford, S., 1992, 'Feminist change in a patriarchal organisation: the experiences of women's initiatives in local government and implications for feminist perspectives on state institutions', in M. Savage and A. Witz (eds.), *Gender and Bureaucracy*, Oxford: Blackwell.

Halley, J., 1993, 'Reasoning about sodomy: act and identity in and after *Bowers v. Hardwick*', *Virginia Law Review* 79, 1721–80.

Halvorsen, R., 1998, 'The ambiguity of lesbian and gay marriages: change and continuity in the symbolic order', *Journal of Homosexuality* 35, 207–31.

Hartmann, H., 1979, 'Capitalism, patriarchy and job segregation by sex', in Z. Eisenstein (ed.), *Capitalist Patriarchy and the Case for Socialist Feminism*, New York: Monthly Review Press.

Hartmann, H., 1981, 'The unhappy marriage of Marxism and feminism: towards a more progressive union', in L. Sargent (ed.), *The Unhappy Marriage of Marxism and Feminism*, London: Pluto Press.

Hartsock, N., 1979, 'Feminist theory and the development of revolutionary strategy', in Z. Eisenstein (ed.), *Capitalist Patriarchy and the Case for Socialist Feminism*, New York: Monthly Review Press.

Hartsock, N., 1990, 'Foucault on power: a theory for women?' in L. Nicholson (ed.), *Feminism/Postmodernism*, London: Routledge.

Harvey, D., 1993, 'Class relations, social justice and the politics of difference', in J. Squires (ed.), *Principled Positions: Postmodernism and the Rediscovery of Value*, London: Lawrence & Wishart.

Hatcher, R., 1994, 'Market relationships and the management of teachers', *British Journal of Sociology of Education* 15, 41–61.

Hatcher, R., 2000, 'Profit and power: business and education action zones', *Education Review* 13, 71–7.

Heller, A. and Fehér, F., 1988, *The Postmodern Condition*, Cambridge: Polity Press.

Hennessy, R., 2000, *Profit and Pleasure: Sexual Identities in Late Capitalism*, London: Routledge.

Herman, D., 1990, 'Are we family? Lesbian rights and women's liberation', *Osgoode Hall Law Journal* 28, 789–815.

Herman, D., 1997, *The Antigay Agenda*, Chicago: University of Chicago Press.

Himmelweit, S., 1995, 'The discovery of "unpaid work": the social consequences of the expansion of "work"', *Feminist Economics* 1, 1–19.

Hird, M., 2000, 'Gender's nature: intersexuality, transsexualism and the "sex"/"gender" binary', *Feminist Theory* 1, 347–64.

Hirsch, A. von, 1986, 'Injury and exasperation: an examination of harm to others and offense to others', *Michigan Law Review* 84, 700–14.

Hirschmann, N., 1996, 'Revisioning freedom: relationship, context, and the politics of empowerment', in N. Hirschmann and C. Di Stefano (eds.), *Revisioning the Political*, Boulder, CO: Westview Press.

Hoggett, P., 1996, 'New modes of control in the public service', *Public Administration* 74, 9–32.

Hollister, J., 1999, 'A highway rest area as a socially reproducible site', in W. Leap (ed.), *Public Sex/Gay Space*, New York: Columbia University Press.

Honig, B. 1996. 'Difference, dilemmas, and the politics of home', in S. Benhabib (ed.), *Democracy and Difference*, Princeton: Princeton University Press.

hooks, b., 1990, *Yearning: Race, Gender, and Cultural Politics*, Boston: South End Press.

Huckman, L. and Hill, T., 1994, 'Local management of schools: rationality and decision-making in the employment of teachers', *Oxford Review of Education* 20, 185–97.

Hughes, D., 2000, 'The use of the possessory and other powers of local authority landlords as a means of social control, its legitimacy and some other problems', *Anglo-American Law Review* 29, 167–203.

Humphrey, J., 1999, 'Organizing sexualities, organizing inequalities: lesbians and gay men in public service occupations', *Gender, Work and Organization* 6, 134–51.

Hunt, M., 1990, 'The de-eroticization of women's liberation: social purity movements and the revolutionary feminism of Sheila Jeffreys', *Feminist Review* 34, 23–46.

Hunter, C. and Nixon, J., 2001, 'Social landlords' response to neighbour nuisance and anti-social behaviour: from the negligible to the holistic?', *Local Government Studies* 27, 89–104.

Hunter, C., et al., 2000, *Neighbour Nuisance, Social Landlords and the Law*, Coventry: JRF/CIH.

Ingold, T., 1993, 'The temporality of the landscape', *World Archaeology* 25, 152–74.

Iyer, N., 1993–4, 'Categorical denials: equality rights and the shaping of social identity', *Queen's Law Journal* 19, 179–207.

Jeffreys, S., 1990, *Anticlimax: A Feminist Perspective on the Sexual Revolution*, London: Women's Press.

Jenkins, R., 1992, *Pierre Bourdieu*, London: Routledge.

Johnson, R., 1986, 'Alice through the fence: Greenham women and the law', in J. Dewar et al. (eds.), *Nuclear Weapons, the Peace Movement and the Law*, Basingstoke: Macmillan.

Johnston, L. and Valentine, G., 1995, 'Wherever I lay my girlfriend, that's my home: the performance and surveillance of lesbian identities in domestic environments', in D. Bell and G. Valentine (eds.), *Mapping Desire*, London: Routledge.

Joseph, G., 1986, 'The incompatible menage à trois: Marxism, feminism, and racism', in L. Sargent (ed.), *The Unhappy Marriage of Marxism and Feminism*, London: Pluto Press.

Judd, D., 1995, 'The rise of the new walled cities', in H. Liggett and D. C. Perry (eds.), *Spatial Practices*, London: Sage.

Kanter, R. M., 1972, *Commitment and Community: Communes and Utopias in Sociological Perspective*, Cambridge, MA: Harvard University Press.

Kaplan, M., 1994, 'Intimacy and equality: the question of lesbian and gay marriage', *Philosophical Forum* 25, 333–60.

Kaplan, M., 1997, *Sexual Justice: Democratic Citizenship and the Politics of Desire*, New York: Routledge.

Kapp, S., 2000, 'Pathways to prison: life histories of former clients of the child welfare and juvenile justice systems', *Journal of Sociology and Social Welfare* 27, 63–74.

Kappel, K., 1997, 'Equality, priority, time', *Utilitas* 9, 203–25.

Kaveney, R., 1999, 'Talking transgender politics', in K. More and S. Whittle (eds.), *Reclaiming Genders: Transsexual Grammars at the Fin de Siècle*, London: Cassell.

Kelling, G. and Coles, C., 1996, *Fixing Broken Windows*, New York: Touchstone.

Kendall, G. and Wickham, G., 2001, *Understanding Culture: Cultural Studies, Order and Ordering*, London: Sage.

Kingdom, E., 1999, 'Citizenship and democracy: feminist politics of citizenship and radical democratic politics', in S. Millns and N. Whitty (eds.), *Feminist Perspectives on Public Law*, London: Cavendish.

Kirk, G., 1989, 'Our Greenham Common: not just a place but a movement', in A. Harris and Y. King (eds.), *Rocking the Ship of State: Towards a Feminist Peace Politics*, London: Westview Press.

Kraatz, M. and Zajac, E., 1996, 'Exploring the limits of the new institutionalism: the causes and consequences of illegitimate organizational change', *American Sociological Review* 61, 812–36.

Kramer, J., 1995, 'Bachelor farmers and spinsters: gay and lesbian identities and communities in rural North Dakota', in D. Bell and G. Valentine (eds), *Mapping Desire*, London: Routledge.

Krasniewicz, L., 1992, *Nuclear Summer: The Clash of Communities at the Seneca Women's Peace Encampment*, Ithaca, NY: Cornell University Press.

Kymlicka, W., 1989, *Liberalism, Community and Culture*, Oxford: Clarendon Press.

Kymlicka, W., 1995, *Multicultural Citizenship: A Liberal Theory of Minority Rights*, Oxford: Clarendon Press.

Kymlicka, W. and Norman, W., 2000, 'Citizenship in culturally diverse societies: issues, contexts, concepts', in W. Kymlicka and W. Norman (eds.), *Citizenship in Diverse Societies*, Oxford: Oxford University Press.

Lacey, N., 1998, *Unspeakable Subjects: Feminist Essays in Legal and Social Theory*, Oxford: Hart.

Laclau, E. and Mouffe, C., 2001, *Hegemony and Socialist Strategy*, London: Verso (2nd edn).

Lacombe, D., 1994, *Blue Politics: Pornography and the Law in the Age of Feminism*, Toronto: University of Toronto Press.

Lansley, S. et al., 1989, *Councils in Conflict: The Rise and Fall of the Municipal Left*, Basingstoke: Macmillan.

Lawrence, P. and Lorsch, J., 1967, *Organization and Environment: Managing Differentiation and Integration*, Boston, MA: Harvard University Press.

Lee, M., and Burrell, R., 2002, 'Liability for the escape of GM seeds: pursuing the victim', *Modern Law Review* 65, 517–37.

Lee, R., 1996, 'Moral money? LETS and the social construction of local economic geographies in Southeast England', *Environment and Planning A* 28, 1377–94.

Lees, S., 2003, 'Making a place for the diasporic community: geography, genealogy and the eruv', presented at the School for American Research, Santa Fe, New Mexico, April 2003.

Liddington, J., 1989, *The Long Road to Greenham*, London: Virago.

Linton, R., 1989, 'Seneca women's peace camp: shape of things to come', in A. Harris and Y. King (eds.), *Rocking the Ship of State: Towards a Feminist Peace Politics*, London: Westview Press.

Livingston, D., 1997, 'Police discretion and the quality of life in public places: courts, communities, and the new policing', *Columbia Law Review* 97, 551–672.

Lorde, A., 1984, *Sister Outsider*, New York: Crossing Press.

Loughlin, M., 1996, *Legality and Locality: The Role of Law in Central–Local Government Relations*, Oxford: Clarendon Press.

Lowndes, V., 1996, 'Varieties of new institutionalism: a critical appraisal', *Public Administration* 74, 181–97.

Lowndes, V., 1997, 'Change in public service management: new institutions and new managerial regimes', *Local Government Studies* 23, 42–66.

Lushington, M., 2000, 'New Labour's secret garden: the privatisation of public education', *Education Review* 13, 59–64.

Mackay, F., 2001, *Love and Politics: Women Politicians and the Ethics of Care*, London: Continuum.

MacKinnon, C., 1983, 'Feminism, Marxism, method, and the state: toward a feminist jurisprudence', *Signs* 8, 635–58.

MacKinnon, C., 1991, 'Reflections on sex equality under law', *Yale Law Journal* 100, 1281–1328.

MacKinnon, C., 1993, *Only Words*, Cambridge: Harvard University Press.

McLaren, J., 1983, 'Nuisance law and the industrial revolution – some lessons from social history', *Oxford Journal of Legal Studies* 3, 155–221.

MacLaughlin, J., 1998, 'The political geography of anti-Traveller racism in Ireland: the politics of exclusion and the geography of closure', *Political Geography* 17, 417–435.

McNay, L., 1996, 'Michel de Certeau and the ambivalent everyday', *Social Semiotics* 6, 61–81.

March, J. and Olsen, J., 1983, 'The new institutionalism: organizational factors in political life', *American Political Science Review*, 78, 734–749.

March, J. and Olsen, J., 1989, *Rediscovering Institutions: The Organizational Basis of Politics*, New York: Free Press.

Markesinis, B., 1989, 'Negligence, nuisance and affirmative duties of action', *Law Quarterly Review* 105, 104–24.

Marren, E. and Levacic, R., 1994, 'Senior management, classroom teacher and governor responses to local management of schools', *Educational Management and Administration* 22, 39–53.

Matsuda, M., 1993, 'Public response to racist speech: considering the victim's story', in M. Matsuda et al. (eds.), *Words that Wound*, Boulder, CO: Westview Press.

Metcalf, B., 1995, *From Utopian Dreaming to Communal Reality. Co-operative Lifestyles in Australia*, Sydney: University of New South Wales Press.

Metzger, J., 1989, 'The eruv: can government constitutionally permit Jews to build a fictional wall without breaking the wall between church and state', *National Jewish Law Review* 4, 67–92.

Meyer, J. and Rowan, B., 1977, 'Institutionalized organisations: formal structure as myth and ceremony', *American Journal of Sociology* 83, 340–63.

Meyer, J. et al., 1983, 'Institutional and technical sources of organizational structure: explaining the structure of educational organizations', in J. Meyer and W. R. Scott (eds.), *Organizational Environments*, Beverly Hills: Sage.

Mill, J. S., 1929, *On Liberty*, London: Watts & Co.

Millbank, J., 1998, 'If Australian law opened its eyes to lesbian and gay families, what would it see?', *Australian Journal of Family Law* 12, 99–139.

Mitchell, D., 1997, 'The annihilation of space by law: the roots and implications of anti-homeless laws', *Antipode* 29, 303–35.

Mohr, R., 1997, 'The case for gay marriage', in R. Baird and S. Rosenbaum (eds.), *Same-Sex Marriage. The Moral and Legal Debate*, Amherst, NY: Prometheus Books.

Monro, S., 2001, 'Transgender Politics', Ph.D. dissertation, Sheffield University.

Moraga, C., 1983, 'Preface', in C. Moraga and G. Anzaldúa (eds.), *This Bridge Called My Back*, New York: Kitchen Table; Women of Color Press.

Moran, L. and Skeggs, B. with Tyrer, P. and Corteen, K., 2004, *Sexuality and the Politics of Violence and Safety*, London: Routledge.

Moruzzi, N., 1994, 'A problem with headscarves: contemporary complexities of political and social identity', *Political Theory* 22, 653–72.

Mouffe, C., 1992, 'Democratic citizenship and the political community', in Chantal Mouffe (ed.), *Dimensions of Radical Democracy*, London: Verso.

Mouffe, C., 1996a, *The Return of the Political*, London: Verso.

Mouffe, C., 1996b, 'Radical democracy or liberal democracy', in D. Trend (ed.), *Radical Democracy*, New York: Routledge.

Mouffe, C., 1996c, 'On the itineraries of democracy: an interview with Chantal Mouffe', *Studies in Political Economy* 49, 131–48.

Mouffe, C., 2000, *The Democratic Paradox*, London: Verso.

Murray, S., 1999, 'Self size and observable sex', in W. Leap (ed.), *Public Sex/Gay Space*, New York: Columbia University Press.

Myslik, W., 1996, 'Renegotiating the social/sexual identities of places: gay communities as safe havens or sites of resistance?' in N. Duncan (ed.), *BodySpace: Destabilizing Geographies of Gender and Sexuality*, London: Routledge.

Naffine, N., 1997, 'The body bag', in N. Naffine and R. Owens (eds.), *Sexing the Subject of Law*, Sydney: LBC Information Services.

Nagel, T., 1998, 'Concealment and exposure', *Philosophy and Public Affairs* 27, 3–30.

Namaste, K., 1996, ' "Tragic misreadings": queer theory's erasure of transgender subjectivity', in B. Beemyn and M. Eliason (eds.), *Queer Studies: A Lesbian, Gay, Bisexual, and Transgender Anthology*, New York: New York University Press.

Namaste, V., 2000, *Invisible Lives: The Erasure of Transsexual and Transgendered People*, Chicago: University of Chicago Press.

Nedelsky, J., 1991, 'Law, boundaries and the bounded self', in R. Post (ed.), *Law and the Order of Culture*, Berkeley: University of California Press.

Neill, A. S., 1968, *Summerhill*, Harmondsworth: Pelican.

Newark, F., 1949, 'The boundaries of nuisance', *Law Quarterly Review* 65, 480–90.

Newman, J., 1995, 'Gender and cultural change', in C. Itzin and J. Newman (eds.), *Gender, Culture and Organisational Change*, London: Routledge.

Newton, T., 1994, 'Discourse and agency: the example of personnel psychology and assessment centres', *Organization Studies* 15, 879–902.

Nixon, J. et al., 1997, 'Towards a learning profession: changing codes of occupational practice within the new management of education', *British Journal of Education* 18, 5–27.

North, P., 1999, 'Explorations in heterotopia: local exchange trading systems (LETS) and the micropolitics of money and livelihood', *Environment and Planning D* 17, 69–86.

O'Doherty, R. et al., 1999, 'Local exchange and trading schemes: a useful strand of community economic development policy?' *Environment and Planning A* 31, 1639–53.

O'Donovan, K., 1993, 'Marriage: a sacred or profane love machine?', *Feminist Legal Studies* 1, 75–90.

Oerton, S., 1997, '"Queer housewives?"': Some problems in theorising the division of domestic labour in lesbian and gay households', *Women's Studies International Forum* 20, 421–30.

Office of the Deputy Prime Minister, 2002, *Living Places: Cleaner, Safer, Greener*, London: HMSO.

Okely, J., 1983, *The Traveller-Gypsies*, Cambridge: Cambridge University Press.

Okin, S., 1994, 'Gender inequality and cultural differences', *Political Theory* 22.

Okin, S., 1998, 'Feminism and multiculturalism: some tensions', *Ethics* 108, 661–84.

Okin, S., 1999, 'Is multiculturalism bad for women?', in S. Okin (ed.), *Is Multiculturalism Bad for Women*, Princeton: Princeton University Press.

Oliver, C., 1992, 'The antecedents of deinstitutionalization', *Organization Studies*, 13, 563–88.

Onlywomen Press (eds.), 1983, *Breaking the Peace*, London: Onlywomen Press.

Pacione, M., 1997, 'Local exchange trading systems as a response to the globalisation of capitalism', *Urban Studies* 34, 1179–99.

Parekh, B., 2000. *Rethinking Multiculturalism: Cultural Diversity and Political Theory*, Basingstoke: Palgrave.

Parpworth, N., 1995, 'The Criminal Justice and Public Order Act 1994: "rave on"?', *Environmental Law and Management* 7, 77–82.

Pateman, C., 1988, *The Sexual Contract*, Cambridge: Polity.

Pepper, D., 1991, *Communes and the Green Vision*, London: Merlin Press.

Phelan, S., 1989, *Identity Politics: Lesbian Feminism and the Limits of Community*, Philadelphia: Temple University Press.

Phelan, S., 1994, *Getting Specific: Postmodern Lesbian Politics*, Minneapolis: University of Minnesota Press.

Phillips, A., 1999, *Which Equalities Matter?* Cambridge: Polity.

Phillips, A., 2003, 'Defending equality of outcome', *Journal of Political Philosophy* 12: 1–19.

Phillips, R. et al., 2000, *Decentring Sexualities: Politics and Representation Beyond the Metropolis*, London: Routledge.

Pierce, C., 1995, 'Gay marriage', *Journal of Social Philosophy* 26, 5–16.

Poland, B., 2000, 'The "considerate" smoker in public space: the micro-politics and political economy of "doing the right thing"', *Health and Place* 6, 1–14.

Pollert, A., 1996, 'Gender and class revisited; or, the poverty of "patriarchy"', *Sociology* 30, 639–59.

Powell, W. and DiMaggio, P. (eds.), 1991, *The New Institutionalism in Organizational Analysis*, Chicago: University of Chicago Press.

Pratt, M., 1995, *S/he*, Ithaca, NY: Firebrand Books.

Purdue, D. et al., 1997, 'DIY culture and extended milieux: LETS, veggie boxes
and festivals', *Sociological Review* 45, 645–67.

Quilley, S., 1997, 'Constructing Manchester's "new urban village": gay space
in the entrepreneurial city', in G. Ingram et al. (eds.), *Queers in Space*,
Seattle: Bay Press.

Raymond, J., 1982, *The Transsexual Empire*, London: Women's Press.

Razack, S. 1998, 'Race, space and prostitution: the making of the bourgeois
subject', *Canadian Journal of Women and the Law* 10, 338–76.

Razack, S., 1999, *Looking White People in the Eye: Gender, Race, and Culture in Courtrooms and Classrooms*, Toronto: University of Toronto
Press.

Rigby, A., 1974, *Communes in Britain*, London: Routledge & Kegan Paul.

Romero, L., 1997, *Home Fronts: Domesticity and its Critics in the Antebellum
United States*, Durham, NC: Duke University Press.

Roseneil, S., 1995, *Disarming Patriarchy: Feminism and Political Action at
Greenham*, Buckingham: Open University Press.

Roseneil, S., 2000, *Common Women, Uncommon Practices: The Queer Feminisms of Greenham*, London: Cassell.

Rothenberg, T., 1995 '"And she told two friends": lesbians creating urban
social space', in D. Bell and G. Valentine (eds.), *Mapping Desire*, London:
Routledge.

Rubin, G., 1989, 'Thinking sex: notes for a radical theory of the politics of
sexuality', in C. Vance (ed.), *Pleasure and Danger: Exploring Female
Sexuality*, London: Pandora Press.

Rubin, H., 1999, 'Trans studies: between a metaphysics of presence and absence', in K. More and S. Whittle (eds.), *Reclaiming Genders: Transsexual
Grammars at the Fin de Siècle*, London: Cassell.

Santos, B. de Sousa, 1987, 'Law: A map of misreading. Toward a postmodern
conception of law', *Journal of Law and Society* 14, 279–302.

Sargisson, L., 2000, *Utopian Bodies and the Politics of Transgression*, London:
Routledge.

Scott, W. R., 1983, 'The organization of environments: network, cultural, and
historical elements', in J. Meyer and W. R. Scott (eds.), *Organizational
Environments*, Beverly Hills: Sage.

Scott, W. R. and Meyer, J., 1983, 'The organization of societal sectors', in
J. Meyer and W. R. Scott (eds.), *Organizational Environments*, Beverly
Hills: Sage.

Sedgwick, E. K., 1990, *Epistemology of the Closet*, Berkeley: University of California Press.

Seel, B., 1997, 'Strategies of resistance at the Pollok Free State road protest camp', *Environmental Politics* 6, 108–39.

Segal, L. and McIntosh, M. (eds.), 1992, *Sex Exposed: Sexuality and the Pornography Debate*, London: Virago Press.

Seidman, S., 1997, *Difference Troubles: Queering Social Theory and Sexual Politics*, Cambridge: Cambridge University Press.

Seidman, S., 2002, *Beyond the Closet: The Transformation of Gay and Lesbian Life*, New York: Routledge.

Seyfang, G., 2001, 'Working for the Fenland dollar: an evaluation of local exchange trading systems as an informal employment strategy to tackle social exclusion', *Work, Employment and Society* 15, 581–93.

Shapiro, J., 1996, *Shakespeare and the Jews*, New York: Columbia University Press.

Sharot, S., 1991, 'Judaism and the secularization debate', *Sociological Analysis* 52, 255–75.

Sharpe, A., 1997, 'Anglo-Australian judicial approaches to transsexuality: discontinuities, continuities and wider issues at stake', *Social and Legal Studies* 6, 23–50.

Sharpe, A., 1999, 'Transgender performance and the discriminating gaze: a critique of anti-discrimination regulatory regimes', *Social and Legal Studies* 8, 5–24.

Shenker, B., 1986, *Intentional Communities: Ideology and Alienation in Communal Societies*, London: Routledge & Kegan Paul.

Shildrick, M., 1997, *Leaky Bodies and Boundaries: Feminism, Postmodernism and (Bio)Ethics*, London: Routledge.

Shor, R. et al., 1980, 'The distinction between the antismoking and nonsmokers' rights movement', *The Journal of Psychology* 106, 129–46.

Shughart, W. and Tollison, R., 1986, 'Smokers versus nonsmokers', in R. Tollison (ed.), *Smoking and Society: Toward a More Balanced Assessment*, Lexington, MA: D. C. Heath & Co.

Simester, A. and von Hirsch, A., 2002, 'Rethinking the offense principle', *Legal Theory* 8, 269–95.

Simons, J., 1985, *Foucault and the Political*, London: Routledge.

Skogan, W., 1990, *Disorder and Decline: Crime and the Spiral of Decay in American Neighborhoods*, New York: Free Press.

Smart, C., 1992, 'The woman of legal discourse', *Social and Legal Studies* 1, 29–41.

Smart, C., 1995, *Law, Crime and Sexuality*, London: Sage.

Smith, A. M., 1998, *Laclau and Mouffe: The Radical Democratic Imaginary*, London: Routledge.

Søland, B., 1998, 'A queer nation? The passage of the gay and lesbian partnership legislation in Denmark, 1989', *Social Politics* 5, 48–69.

Sommella, L. / Wolfe, M., 1997, 'This is about people dying: the tactics of early ACT UP and Lesbian Avengers in New York City', in G. Ingram et al. (eds.), *Queers in Space*, Seattle: Bay Press.

Spelman, E., 1990, *Inessential Woman: Problems of Exclusion in Feminist Thought*, London: The Women's Press.

Spencer, J., 1989, 'Public nuisance – a critical examination', *Cambridge Law Journal* 48, 55–84.

Spiro, M., 1963, *Kibbutz: Venture in Utopia*, New York: Schocken Books.

Squires, J. (ed.), 1993, *Principled Positions: Postmodernism and the Rediscovery of Value*, London: Lawrence & Wishart.

Steele, J., 1995, 'Private nuisance and the environment: nuisance in context', *Legal Studies* 15, 236–59.

Steinberger, P., 1999, 'Public and private', *Political Studies* 47, 292–313.

Stone, S., 1991, 'The empire strikes back: a posttranssexual manifesto', in J. Epstein and K. Straub (eds.), *Body Guards*, New York: Routledge.

Stychin, C., 1994, 'A postmodern constitutionalism: equality rights, identity politics, and the Canadian national imagination', *Dalhousie Law Journal*, 17, 61–82.

Stychin, C., 1995, *Law's Desire: Sexuality and the Limits of Justice*, London: Routledge.

Stychin, C., 1996, 'To take him at his "word": theorizing law, sexuality and the US military exclusion policy', *Social and Legal Studies* 5, 179–200.

Stychin, C., 1999, 'New Labour, New "Britain"? Constitutionalism, sovereignty, and nation/state in transition', *Studies in Law, Politics and Society* 19, 139–64.

Stychin, C., 2003, *Governing Sexuality: The Changing Politics of Citizenship and Law Reform*, Oxford: Hart.

Sullivan, A., 1997, 'Virtually normal', in R. Baird and S. Rosenbaum (eds.), *Same-Sex Marriage. The Moral and Legal Debate*, Amherst, NY: Prometheus Books.

Taylor, C., 1992, *Multiculturalism and the "Politics of Recognition"*, Princeton: Princeton University Press.

Tessler, R. et al., 2001, 'Gender differences in self-reported reasons for homelessness', *Journal of Social Distress and the Homeless* 10, 243–54.

Thomson, M., 2001, 'eXistenZ: Bio-ports/ boundaries/ bodies', *Legal Studies* 21, 325–43.

Thorne, L., 1996, 'Local exchange trading systems in the United Kingdom: a case of re-embedding', *Environment and Planning A* 28, 1361–76.

Tobin, A., 1990, 'Lesbianism and the Labour Party: the GLC experience', *Feminist Review* 34, 56–66.

Trachtman, J., 1983, 'Pornography, padlocks and prior restraints: the constitutional limits of the nuisance power', *New York University Law Review* 58, 1478–529.

Trosch, W., 1993, 'The third generation of loitering laws goes to court: do laws that criminalize "loitering with the intent to sell drugs" pass constitutional muster?', *North Carolina Law Review* 71, 513–68.

Troyer, R. and Markle, G., 1983, *Cigarettes: The Battle over Smoking*, New Brunswick: Rutgers University Press.

Tuggle, J. and Holmes, M., 1997, 'Blowing smoke: status politics and the Shasta County smoking ban', *Deviant Behavior* 18, 77–93.

Tyner, J. and Houston, D., 2000, 'Controlling bodies: the punishment of multiracialized sexual relations', *Antipode* 32, 387–409.

Uyl, D., 1986, 'Smoking, human rights and civil liberties', in R. Tollison (ed.), *Smoking and Society: Toward a More Balanced Assessment*, Lexington, Mass: D.C. Heath & Co.

Valentine, G., 1993, '(Hetero)sexing space: lesbian perceptions and experiences of everyday spaces', *Environment and Planning D* 11, 395–413.

Valentine, G., 1995, 'Out and about: geographies of lesbian landscapes', *International Journal of Urban and Regional Research* 19, 96–111.

Valentine, G., 1996, '(Re)negotiating the "heterosexual street": lesbian productions of space', in N. Duncan (ed.), *BodySpace*, London: Routledge.

Valins, O., 2000, 'Institutionalised religion: sacred texts and Jewish spatial practice', *Geoforum* 31, 575–86.

Valverde, M., 1985, *Sex, Power and Pleasure*, Toronto: Women's Press.

Vance, C. (ed.), 1989, *Pleasure and Danger: Exploring Female Sexuality*, London: Pandora Press.

Vincent, P. and Warf, B., 2002, 'Eruvim: Talmudic places in a postmodern world', *Transactions of the Institute of British Geographers* 27, 30–51.

Vogel, L., 1983, *Marxism and the Oppression of Women: Toward a Unitary Theory*, London: Pluto Press.

Vorspan, R., 1998, 'The political power of nuisance law: labor picketing and the courts in modern England, 1871–present', *Buffalo Law Review* 46, 593–703.

Walby, S., 1990, *Theorizing Patriarchy*, Oxford: Blackwell.

Waldron, J., 1991, 'Homelessness and the issue of freedom', *University of California Law Review* 39, 295–324.

Waldron, J., 1995, 'Minority cultures and the cosmopolitan alternative', in W. Kymlicka (ed.), *The Rights of Minority Cultures*, Oxford: Oxford University Press.

Walzer, M., 1984, 'Liberalism and the art of separation', *Political Theory* 12, 315–30.

Walzer, M., 1986, 'Pleasures and costs of urbanity', *Dissent* 33, 470–5.

Ward, C., 1997, 'On difference and equality', *Legal Theory* 3, 65–99.

Warner, M., 1999, *The Trouble with Normal*, New York: Free Press.

Wayland, S., 1997, 'Religious expression in public schools: *kirpans* in Canada, *hijab* in France', *Ethnic and Racial Studies* 20, 545–61.

Weeks, J., 1993, 'Rediscovering values' in J. Squires (ed.), *Principled Positions: Postmodernism and the Rediscovery of Value*, London: Lawrence and Wishart.

Weeks, J., 1995, *Invented Moralities: Sexual Values in an Age of Uncertainty*, Cambridge: Polity Press.

Weeks, J. et al., 1999, 'Partners by choice: equality, power and commitment in non-heterosexual relationships', in G. Allan (ed.), *The Sociology of the Family*, Oxford: Blackwell.

Weeks, J. et al., 2001, *Same Sex Intimacies: Families of Choice and Other Life Experiments*, London: Routledge.

Weitzman, B. et al., 1990, 'Pathways to homelessness among New York City families', *Journal of Social Issues* 46, 125–40.

Welsh, I. and McLeish, P., 1996, 'The European road to nowhere: anarchism and direct action against the UK roads programme', *Anarchist Studies* 4, 27–44.

Werbner, P., 1996, 'Fun spaces: on identity and social empowerment among British Pakistanis', *Theory, Culture and Society* 13, 53–79.

West, C. and Fenstermaker, S., 2002, 'Doing difference', in S. Fenstermaker and C. West (eds.), *Doing Gender, Doing Difference: Inequality, Power, and Institutional Change*, New York: Routledge.

West, C. and Zimmerman, D., 2002, 'Doing gender', in S. Fenstermaker and C. West (eds.), *Doing Gender, Doing Difference: Inequality, Power, and Institutional Change*, New York: Routledge.

Weston, K., 1995, 'Forever is a long time: romancing the real in gay kinship ideologies', in S. Yanagisako and C. Delaney (eds.), *Naturalizing Power: Essays in Feminist Cultural Analysis*, New York: Routledge.

Whittington, R., 1992, 'Putting Giddens into action: social systems and managerial agency', *Journal of Management Studies* 29, 693–712.

Whittle, S., 1996, 'Gender fucking or fucking gender? Current cultural contributions to theories of gender blending', in R. Ekins and D. King (eds.), *Blending Genders*, London: Routledge.

Whittle, S., 1998, 'The trans-cyberian mail way', *Social and Legal Studies* 7, 389–408.

Wightman, J., 1998, 'Nuisance – the environmental tort? *Hunter* v. *Canary Wharf* in the House of Lords', *Modern Law Review* 61, 870–85.

Wilchins, R. A., 1997, *Read My Lips: Sexual Subversion and the End of Gender*, Ithaca, NY: Firebrand Books.

Williams, C., 1996, 'The new barter economy: an appraisal of local exchange and trading systems (LETS)', *Journal of Public Policy* 16, 85–101.

Wintemute, R. and Andenæs, 2001, *Legal Recognition of Same-Sex Partnerships*, Oxford: Hart.

Winter, T., 2000, 'Reclaiming quiet enjoyment', *Adviser* 80, 9–11.

Wolfson, E., 1996, 'Why we should fight for the freedom to marry: the challenges and opportunities that will follow a win in Hawaii', *Journal of Gay, Lesbian, and Bisexual Identity* 1, 79–89.

Yar, M., 2001, 'Beyond Nancy Fraser's "perspectival dualism"', *Economy and Society* 3, 288–303.

Young, I., 1986, 'Beyond the unhappy marriage: a critique of the dual systems theory', in L. Sargent (ed.), *The Unhappy Marriage of Marxism and Feminism*, London: Pluto Press.

Young, I. M., 1990a, *Justice and the Politics of Difference*, Princeton: Princeton University Press.

Young, I. M., 1990b, *Throwing Like a Girl and Other Essays in Feminist Philosophy and Social Theory*, Bloomington: Indiana University Press.

Young, I. M., 1993, 'Together in difference: transforming the logic of group political conflict', in J. Squires (ed.), *Principled Positions: Postmodernism and the Rediscovery of Value*, London: Lawrence & Wishart.

Young, I. M., 1994, 'Gender as seriality: thinking about women as a social collective', *Signs* 19, 713–38.

Young, I. M., 1997a, 'Unruly categories: a critique of Nancy Fraser's dual systems theory', *New Left Review* 222, 147–60.

Young, I. M., 1997b, *Intersecting Voices: Dilemmas of Gender, Political Philosophy and Policy*, Princeton: Princeton University Press.

Young, I. M., 2000, *Inclusion and Democracy*, Oxford: Oxford University Press.

Index

Adams, Tracey 45
agency
  of organisational actors 153, 156–157
  restructured 159–160
akinship 100, 101
  and same-sex marriage 110–111
Aldridge, Theresa 175
anti-Semitism 30
antisocial behaviour (*see also* nuisance) 120,
  123, 129–130, 137–138, 140, 141
Appleton, Matthew 185, 187
axes model (*see also* social inequality) 44,
  47–49

Bauman, Zygmunt 13
Berlin, Isaiah 121
Bey, Hakim 166
Bickford, Susan 67
binaries of oppression 52
Boga, Terence 130
boundaries
  community 167–168, 178–179, 187–189
  permeability 183–185, 206
  Summerhill School 181
  through space 172
boundary
  actors 188–189
  techniques 167–168
Bourdieu, Pierre 144, 151
Boyd, Susan 112
Bradford Corporation v. *Pickles* 123
buffering 162
Butler, Judith 20–21, 33–34, 71, 144
Butler, Sandra 146

capitalism 55–56
Card, Claudia 111
Carrington, Christopher 91, 112–113, 115, 117

Carroll, William 169
choice 73
cigarette smoking 60–65
class 43–44, 45, 55–56, 58
classification of gender 86–88
Combahee River Collective 44
common sense of nuisance 138–139
communities
  and counter-normative pathways 203–204
  maintenance of 172–173
  as nuisance 136
community
  boundaries 167–168, 178–179, 187–189
  challenge of 207
  and diversity 197
  impact of 205–206
  and sameness 186–187
Conaghan, Joanne 120
counter-hegemony
  governance technologies of 199–200
counter-normative
  age 180–181
  effects 184–185
  nuisance discourse 139
  pathways 150–151, 154–157, 161, 170–171,
    203–204
  principles 95
  proper place 128–129
  routines 143–144
  survival 167
Crenshaw, Kimberle 45–46, 70
crisis/overload 156
Croall, Jonathan 175, 177, 190
cultural associations, gender 130–132
cultural harm 191–192

de Certeau, Michel 146–147
democracy at Summerhill School 180

difference (*see also* diversity, social
     inequality) 19–21, 41–42
dilemmatic environment 162–163, 197,
     204
disability 73
disruption as politics 135–136
diversity
     community 197
     of pathways 204
     politics 7–8, 15–37
domestication 56–58
dual-systems theory 43–44
Dumm, Thomas 24
Dunn, James 146
Dworkin, Ronald 72–74, 90

Edensor, Tim 146
education policy 159
Education Reform Act 1988 159
Eisenstein, Zillah 15, 44
environmental factors 155–156
environmental messiness 161–162
equal opportunity policies 154–161
equal recognition 74–77
equality
     and communities 186
     and diversity 196–197
     of gender 83–88
     of individuals 70
     and LETS 176
     and norms 92
     of power 77–83
     of resources 71–74
eruv 16–35
     opponents 16–17

family 112
     as nuisance 126
feminism (*see also* gender) 43–47,
     56–57
Foucault, Michel 52, 78–79, 82
Fraser, Nancy 6, 20, 40–43, 71–72, 74, 75,
     193, 198
freedom 23–27, 36, 121–126
     and age 180
     and diversity 186–187
     positive 138
Freeman, Jo 204
Fridman, G. 123

Galeotti, Anna 31, 75, 192, 200
gender (*see also* feminism) 43–44, 45, 56–59
     bodies 130–131
     equality of 83–88
     morphing 85–86
Gewirtz, Sharon 160
governance, sexuality 97–98
governmentality 185

Greenham Common women's peace camp
     169–173, 183–184, 206
group
     equality of 69–70
     lesbian and gay 102
group-based perspective 46–47, 49–50

harm 29–35, 37, 201–202
     and equality of power 82
     and evaluation 194
     values underlying 118–119
Heller, Agnes 19, 20
homosexuality (*see also* lesbian and gay)
     52–53
     governance of 104
Honig, Bonnie 162, 163
Hunt, Margaret 19

identity 19–21, 35–36
     and being 70–71
     dominant 76–77
improper place 134–135
inequality (*see* social inequality)
injury (*see also* harm)
     symbolic 31–35
intersectionality (*see also* social inequality)
     45–46, 48–49
intimate/impersonal (*see also* private,
     public/private divide) 56–59, 129
     and LETS 175
isomorphism 52
     and LETS 174

Jewish communities 17–18, 22–23, 25–29,
     30, 34
Johnson, Rebecca 170–171

Kanter, Rosabeth 187
Kapp, Stephen 145
Kelling, George 122
Kirk, Gwyn 189
Krasniewicz, Louise 171, 172
Kymlicka, Will 5, 23–24, 25, 38

Lacey, Nicola 8
Laclau, Ernesto 92, 116
land boundaries 131–132
law
     discrimination towards smokers 60–63
     education 159
     environmentalise 153, 181–182
     gender classifications 86–88
     nuisance (*see also* antisocial behaviour)
          120, 122–124, 127–128, 131–132
     and pathways 150
     resistance to 170–171
     same-sex partnership recognition 102–114
     at Summerhill School 180, 181–182

lesbian and gay (*see also* homosexuality)
  conservatives 103
  critics of marriage 103
  equality initiatives 155–158
  spousal rights 102–114
Liddington, Jill 173
Linton, Rhoda 170
local exchange trading systems (LETS) 169,
  173–179, 204–205
local government 154–159
loitering 123

MacLaughlin James 128
mainstream, gravitational pull 176–178
marriage
  heterosexual 108
  same-sex 102–114, 151
Matsuda, Mari 33
Mcleish, Phil 135
mental freedom 124–126
Millbank, Jenni 107
minority status (*see also* diversity, social
  inequality) 5–6
Mouffe, Chantal 6, 20, 21, 199, 202

Naffine, Ngaire 130–131
Nagel, Thomas 92, 94
Namaste, Viviane 90
Neill, A. S. 179
non-normative practices 204–205
normative principles 92–96, 201–203
  and inequality 116
norms
  of appropriateness 155–156
  educational 159–161
nuisance (*see also* harm) 118–141, 202
  and freedom 121–126
  and legitimacy 132–133
  public 120
  tort law 120, 127

offence (*see also* harm) 32–33
Okin, Susan 192
oppression (*see also* social inequality) 3,
  42–47
organisational environment 152–161
orthodox Judaism 22

Parekh, Bhikhu 192, 200
pathways
  counter-normative 170–171
  de facto 148–149
  de jure 150
  diversity of 162–163
  inequality 149
  as metaphor 144–148
  resistance 157
  survival of 150–151
  unpredictability 160–161
permeability
  of boundaries 206
  of communities 187–189
Phillips, Anne 83–84
Pierce, Christine 111
pluralism, limits 22
Poland, Blake 62
politics, as playful 173
Pollert, Anna 56
power 77–83
  and nuisance 132–133
Pratt, Minnie Bruce 87
preferences 3–4
privacy 27–29
  (homo)sexuality 100–101
private and same-sex relationships 112–113
proper, respectable gays 114–115
proper place 96–99
  through contract 105–106
  impact on 109–110
  of marriage 108
  and nuisance 126–129
public/private 56, 99–101

race 44, 45, 58
  equality of 76
racism and family 112
Razack, Sherene 82–83
recognition 74–77
  of communities 188
  of same-sex relationships 114
resistance through nuisance 128–129,
  132–133, 135–136
Rigby, Andrew 166–167
Rosenberg v. *City of Outremont* 27–28
Roseneil, Sasha 170, 173, 183, 189
routines
  survival 144
  value of 143–144

same-sex relationships and domestic labour
  113, 115
sameness and equality 196
Santos, Boaventura de Sousa 147
scale as metaphor 147–148
Section 28 Local Government Act 1988
  157–158
Sedgwick, Eve Kosofsky 52–53, 54
Seel, Ben 183
Seidman, Steven 21, 54
self-interest 123
  and LETS 179
sex radicalism 20, 32

sexuality 52–53
  governance 97–98
Sharpe, Andrew 87
Shenker, Barry 168
social inequality 41–43, 191
  axes of power 44, 47–49
  dismantling of 198–199
  distinctiveness 65–66
  and norms 92
  as organising principles 51–55, 194–196
  of pathways 149
  at Summerhill School 182
  undoing 83–88
social dynamics 54–59, 64, 129–130, 195
social logic 148–149, 162
space
  eruv 16–35
  lesbian and gay 98–99
  meaning 34–35
  and LETS 177
speech, harm 33–34
spousal recognition
  entitlements 105
  form 105–107
status and nuisance 127–128
stranger in marriage 111–112
strangers, powerful 113–114
Stychin, Carl 61, 113
Sullivan, Andrew 111
Summerhill School 169, 179–183
  and dispute resolution 205
symbolic activism
  and nuisance 134–135

textual meaning 33
tolerance 18–19
tort law, nuisance 120, 127
trading through LETS 173–179
transgender 149
  and gender politics 84–88, 199
  and nuisance 134
transgressive politics (*see also* counter-
      normative, resistance, symbolic activism)
  around marriage 107–108
travellers 128
Tuggle, Justin 62

Valins, Oliver 26, 38
value 42
values
  as boundaries 178
  circulation of 53–54
  as community maintenance 172–173
  and inequality 195
  of recognition 75–76
  smoking-related 63–64
Vogel, Lise 67

Waldron, Jeremy 122, 128, 129
Ward, Cynthia 196
Weeks, Jeffrey 7, 15, 18–19, 21, 24, 27, 29–30,
      32–33, 104, 105, 111, 199
Weitzman, Beth 145
Whittington, Richard 152
Wilchins, Riki 145–146, 149

Young, Iris Marion 3, 27, 29, 43, 46, 74